Desertification

Nick Middleton

OXFORD UNIVERSITY PRESS

ACKNOWLEDGEMENTS

The publishers and author would like to thank the following people for their permission to use copyright material:

p.4 Mark Edwards, Still Pictures; p.14 David Schultz, Tony Stone Worldwide; p.15 Mark Edwards, Still Pictures; p.17 Mark Edwards, Still Pictures; p.21 *left* E.J. Davis, *right* Steve McCutcheon, both Frank Lane Picture Agency; p.22 Topham Picture Source; p.24 Steve McCurry, Magnum Photos; p.34 NOAA, UNEP; p.39 Stuart Franklin, Magnum Photos.

The cover photograph is reproduced by permission of Mark Edwards, Still Pictures.

Illustrations are by Herb Bowes Graphics, Oxford.

Every effort has been made to trace and contact copyright holders, but this has not always been possible. We apologise for any infringement of copyright.

Oxford University Press, Walton Street, Oxford OX2 6DP

Oxford New York Toronto Delhi Bombay Calcutta Madras Karachi Petaling Jaya Singapore Hong Kong Tokyo Nairobi Dar es Salaam Cape Town Melbourne Auckland and associated companies in Berlin Ibadan

Oxford is a trade mark of Oxford University Press

© Oxford University Press 1991

ISBN 0 19 913367 0

Typeset by Gem Publishing Company, Wallingford
Design and artwork by Herb Bowes Graphics, Oxford
Printed in Great Britain by
M & A Thomson Litho Ltd., East Kilbride, Scotland

PREFACE

Contemporary Issues in Geography is a series of books dealing with issues of concern to today's society. The series was developed as a result of our own teaching needs, especially when preparing INSET courses for teachers in Coventry and Warwickshire.

Hugh Matthews and Ian Foster, Series Editors

This book

Desertification is an appropriate subject for inclusion in this new series on contemporary geographical issues since its study demands the application of all the skills that the geographer is concerned with. It is an issue that is studied at a variety of spatial scales from global through to village level. A full appreciation of time is also required because one of the key difficulties is in trying to distinguish trends in dryland ecosystems which by their very nature are highly dynamic from day to day, from season to season and from decade to decade. Desertification also unites the traditional divide within the subject by looking at the ways in which human society uses the physical environment and how natural systems affect society. This book also takes the reader through a wide range of types of evidence and methods of study used in looking at desertification, and in doing so spotlights some question marks which have made the subject an issue.

This book questions some of the assumptions behind the issue and in the process throws up more problems than it solves. The reason for doing this is not to confuse the reader, but to highlight the very real problems of working on such a broad scale physical and social topic.

Nick Middleton

Note: At the time of writing Kuwait, a country used as a case study in this book, had been invaded by Iraqi military forces. Examples from Kuwait refer to the country pre-August 1990.

CONTENTS

Part I

DESERTIFICATION – A DRYLAND PROBLEM

Most of us have seen distressing pictures of starving Africans, children with distended bellies, stick-like limbs and gaunt faces. We have observed people reduced to scratching in the dust for a grain of millet, to slaughtering their livestock or watching them slowly die. They are refugees from famine, ecological disaster or war, surviving, or perishing, in makeshift camps and eating when they can from foreign food shipments. These heart-rending images of fellow human beings clinging to their final possession, life itself, are usually accompanied by captions or voice-overs that use a familiar vocabulary of words and phrases. These include famine, drought, civil strife, war, poverty, under-development, Third World and desertification.

It is appropriate that desertification, the subject of this book, is used in the same breath as these other words, since it is closely allied to these other complex problems. Although desertification is not exclusively an African or Third World phenomenon, it was in West Africa that the problem was first identified and where it is still the most pressing problem area.

1 WHAT IS DESERTIFICATION?

1.1 Problems of definition

What is desertification? Unfortunately, the answer is not a simple one, since desertification is a rather controversial name for a very complicated subject.

Literally, desertification means the making of a desert. The term was first used in 1949 by a French forester, Aubreville, who was working in West Africa. He used it to describe what he saw was happening in the region: the clearance of forests in humid areas adjacent to drylands which continued until the land was transformed from woodland to desert. Desertification became a more widely-used term after the tragic circumstances in the African Sahel region became daily international news in the 1970s. This event prompted the United Nations to hold a Conference on Desertification (UNCOD) in Nairobi in 1977. The definition of desertification adopted by the conference was:

'the diminution or destruction of the biological potential of the land, that can ultimately lead to desert-like conditions' (United Nations Environment Programme, Desertification—Its Causes and Consequences, Oxford, Pergamon, 1977).

However, many of the scientists working on desertification issues have formed their own definitions. Some incorporate ideas of the causes behind desertification and its consequences. Hence many talk of a loss of total vegetation cover, or a loss of plants that are good for grazing. Some talk of soil loss, others of increased hazards for human occupants of a desertified area.

The idea that human action plays a role in producing desertification is central to the debate. Some authorities believe that it is the major cause, while others advocate that climatic factors are as important if not dominant.

There are a number of human actions that may cause desertification. They include grazing too many animals on a piece of land and cutting vegetation for fuelwood. In both cases this action leaves soil exposed to increased erosion by wind and water.

The main climatic factor that contributes to

desertification is a lack of rainfall. This could be due to a drought or it may result from a long term change to a drier climate.

Some authors do not agree with the term *desertification* at all, preferring to call the phenomenon *desertization*. These authors argue that desertification as defined above is the same as *land degradation* which can happen well away from the world's drylands. They argue that desertification, or desertization, should be confined to arid and semi-arid lands because it is a special form of land degradation which requires specific action to combat it. Desertization has been defined as:

> 'the spread of desert-like conditions in arid or semi-arid areas, due to man's influence or climatic change' (Rapp, A. A review of Desertization in Africa, Stockholm, Secretariat for International Ecology, 1974).

This lack of consensus about the nature of desertification was well-illustrated during a recent conference convened by the European Community entitled *Desertification in Europe*. During the conference several speakers suggested that the conference title was inappropriate since desertification as confined to arid and semi-arid ecosystems is limited to only a few European regions in Spain, Greece and Turkey.

Still further confusion stems from the widespread use of a range of other terms. *Desertification* and *desertization* have been mentioned. Others include *desiccation, aridization, aridification, crypto-desertification, desert encroachment, man-made deserts, desert advance, desert creep, desert expansion* and *spreading deserts*.

All these terms are used to mean a more-or-less similar phenomenon, but there are significant differences within all these terms that have an important influence on just what the scientist sets out to study.

1.2 Wrong definitions produce wrong solutions

A common problem that arises from many of these terms concerns how desertification comes about. The terms *desert encroachment, desert advance, desert creep, desert expansion* and *spreading deserts* all suggest that the main problem of arid areas is the expansion of existing deserts at their edges. Statistics on the rate of such expansion are often used by governments, conservation agencies and politicians. For example, on 14th March 1986, Vice-President Bush was being urged to supply aid to Sudan because 'desertification was advancing at 9 km per annum'. On 11th September 1986, during a debate in the European Parliament on aid to Africa, Winifred Ewing (Scottish Nationalist) urged that aid should be sent to the Sahel, because the desert was 'advancing at 8 km a year'.

These are powerful and evocative statements. They are also very misleading. The image of a sand dune moving from the desert to engulf productive agricultural land is strong, but in reality the most serious problems do not occur at the desert edge. This has been repeatedly stated during UNCOD and since. By far the more common problem is that of patches or islands of degradation, well beyond the desert edge, that may expand and coalesce, eventually perhaps giving the impression of an advancing front (Figure 1.1).

Nevertheless, the statistics are still quoted and plans to combat this so-called advance put into action. Programmes to halt desert advance often entail planting lines of trees at the desert edge, to form *green belts*, or *shelter belts*. The idea is to build a green wall to stop the shifting sands in their tracks. Another commonly employed solution is to plant sand dunes with vegetation in order to stop them moving, as has happened in Algeria, Iran, Sudan, Somalia and China.

In both cases the planting of greenery is a visible sign that something is being done to tackle the problem. But since the expansion of the existing desert at its edge is rarely the main hazard in the desertification process, shelter belts and planted dunes are usually little more than *green irrelevancies*, falsely raising hopes that the problem is being solved.

Problems of definition, therefore, are important because the way in which desertification is defined will influence the identification of the processes operating to cause it. This, in turn, will influence the methods decided upon to combat the problem. So if the definition is wrong, the wrong causes of the problem may be identified and ineffective measures will be employed to solve the problem.

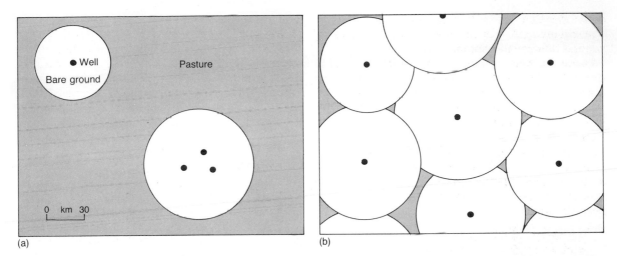

(a) In the original situation pasture is eaten by grazing animals around wells. (b) After a well-digging project large-scale loss of pasture has taken place. More wells has allowed herds to increase and vegetation is lost over a wide area

Figure 1.1 *Typical pattern of desertification by overgrazing*

1.3 An appropriate definition – a range of solutions

The definition of desertification that will be used in this book is as follows:

'Desertification in its least ambiguous form, is the notion that the extent of deserts – dry areas with few plants – is increasing, usually into the semi-arid lands' (Warren, A. and Agnew, C. An Assessment of Desertification and Land Degradation in Arid and Semi-Arid Areas, London, IIED, 1988).

This definition is useful because it is simple, it confines the problem to desert fringes but not their edges, and gives some idea of what is a desert, another problem of definition that will be looked at in Section 2. If we accept this definition we can now examine the range of possible causes of desertification and the appropriate types of solution.

If, for example, the problem is thought to be a long term climatic change to drier conditions then the solution may be to move people out of the area or introduce different methods for using the land. If, however, the problem is thought to be drought lasting a few years, then food aid over that period may be the answer. If vegetation has been lost without any variations in climate then re-seeding or re-planting may be called for. If the problem is one of falling productivity due to over-cultivation, overgrazing or overirrigation then farmers and herders will need help to develop more suitable land practices.

Hence it is preferable to set out with a wide-ranging definition of the desertification problem. With such a definition, the scientist can proceed to study desertification in a particular region, identify the nature, extent and causes of the problem in that region and formulate some form of strategy to combat it.

2 WHERE IS THE PROBLEM?

If we accept the view that desertification occurs at or near existing desert margins and that desertification itself is the spread of desert-like conditions, we need to examine two questions in some detail. First, what is a desert and thus what are desert-like conditions? From the answers to this question we should be able to respond to the second question, where are the desert margins?

2.1 What is a desert?

The following discussion on desert definition will concentrate on the hot and mid-latitude deserts as opposed to the polar deserts. Although polar deserts conform to many of the criteria we shall discuss, they are not generally regarded as areas subject to the desertification hazard.

Most people have an idea of what a desert involves. The name conjures up a number of images such as empty of life, waterless and unproductive. Sand dunes, camels and lack of water are perhaps the most common perceptions. Each has some validity. Sand dunes are typical desert landforms, but they only cover about 25 per cent of world deserts and in some areas, such as the Gobi Desert and the deserts of the US south-west, they are rare features. Camels are animals that are extremely well-adapted to the rigours of the desert environment. Unfortunately, they move, both in and out of deserts and are therefore not terribly useful for our present problem.

A lack of water, or aridity, is perhaps one of the few characteristics that typify all deserts. Various classifications of aridity have been described by geographers and climatologists using annual precipitation statistics.

For example, the United Nations Food and Agriculture Organisation (FAO) uses varying precipitation limits:

arid from 80–150 to 200–350 mm a year, and *semi-arid* from 200–500 to 450–500 mm a year with winter rains but 300–400 to 700–800 with summer rains.

Although annual average precipitation seems a straightforward, easily measured parameter to use when classifying aridity, it has some drawbacks. Most serious is the nature of desert rainfall, which is highly variable both in time and space. A desert area may go for several years with no rainfall at all and then receive a whole *average* year's rainfall in one storm (see Table 1.1). This *interannual variation* can reach 100 per cent in true desert areas (for comparison the interannual variation for a humid temperate climate such as in Western Europe is around 10–20 per cent). The geographical distribution of rainfall in deserts is also important. It is often highly localised. An area of only a few square

Table 1.1 Climate data for drylands (and London)

Station	Average annual precipitation (mm)	Maximum precipitation in 24 hrs (mm)	Average daily temperature (°C)	Average relative humidity (%)
Bilma (Niger)	22	49	26.6	27
Jidda (Saudi Arabia)	76	140	27.7	56
Kashgar (China)	86	25	12.2	58
Antofagasta (Chile)	8	38	16.2	74
Multan (Pakistan)	170	155	26.0	48
Phoenix (USA)	184	78	21.4	43
Alice Springs (Australia)	252	147	20.6	37
London (UK)	593	60	10.7	73

kilometres may receive a downpour while the area all around is unaffected.

With this in mind, the annual average rainfall figure, in areas with such large interannual variations and highly localised storms in zones often with few recording stations, is a poor indicator of desert conditions if used on its own. Another problem with using simple rainfall figures is that it takes no account of another factor which plays a major role in determining aridity: temperature.

Temperature will determine evaporation rates and therefore how much of the precipitation remains on or in the soil as effective moisture. A number of indices have been devised to link these important parameters of rainfall and the intensity of solar radiation received. The index used for the preparation of aridity maps for the 1977 United Nations Conference on Desertification (UNCOD) was that devised by a Soviet scientist, Budyko. Budyko's *dryness ratio* is defined as the number of times the mean net radiation at the earth's surface (R) in a year can evaporate the mean precipitation (P). The amount of heat required to evaporate a unit volume of water is known as the latent heat of vaporisation (L). The dryness ratio (D) is thus calculated as: $D = R/LP$.

Using this index an area was designated a desert if the dryness ratio was more than 10; arid or desert margin if D was between 7 and 10; semi-arid between 2 and 7; and sub-humid if D was less than 2.

This measure of aridity is shown geographically in Figure 1.2. It is just one way that geographers use to define drylands. Others include vegetation, landforms and soils. As one moves from desert to sub-humid regions, for example, vegetation types change from barren wastelands to grasslands with the occasional tree or shrub, to a mix of taller open woodland and grasses, to types of closed forests. Table 1.2 illustrates a number of these characteristics of dryland areas along with the types of use that they can be put to by human populations. This catalogue of considerations and characteristics provides a basic profile of the types of regions where desertification occurs.

Estimates vary on the proportion of the world's land area that is made up of desert, depending on which of the many criteria for classification are adopted. Most are around 30 per cent, indicating that about one third of the world's total land surface is taken up by extremely arid, arid and semi-arid lands. *It is these three zones, plus the sub-humid zone, that are at risk from desertification.*

Extremely arid and arid zones can be called deserts, while semi-arid and sub-humid zones are called the desert margin. But the nature of the desert environment is that its boundaries are dynamic, changing with the seasons and years. Hence it is convenient to talk of the world's *drylands*, which include both the desert and its margins.

2.2 A global problem

Figure 1.3 shows the world distribution of desertification. The map was prepared by the United Nations Environment Programme (UNEP) in their attempt to understand the global pattern of the process. It is based on expert estimates of the extent of the problem on a global scale. Four categories of desertification are used: slight, moderate, severe and very severe. The similarity in pattern between Figures 1.3 and 1.2 is immediately apparent: desertification is happening in the world's drylands. In total, UNEP estimates that desertification threatens 35 per cent of the earth's land surface and 19 per cent of its population. Clearly, according to these figures, desertification is a major world problem.

Table 1.2 Characteristic features of dryland regions

Type	Dryness ratio*	Precipitation (mm)	Precipitation interannual variation*	Vegetation	Land use
Hyper-arid	>10	Very low (<25)	100% or more	Very little or none permanent. Some after rain or dew	Oasis culture. Nomadism
Arid	7–10	From 80–150 to 200–350. Low humidity	50–100%	Sparse. Found in water channels	Pastoralism. No farming unless irrigated
Semi-arid	2–7	From 300–400 to 700–800 with summer rains. From 200–250 to 450–500 with winter rains	25–50%	Savannah or steppe grass. Some thorny shrubs	Rainfed cultivation and sedentary livestock
Sub-humid	<2	Abundant with usually more than six humid months	<25%	Grasses and woodlands	Rainfed cultivation and industrial crops

* Note: these terms are defined in the text

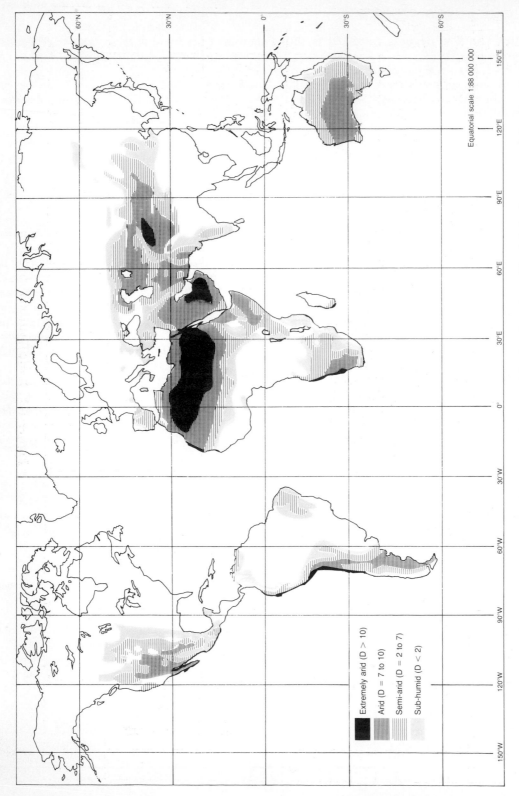

Figure 1.2 *World distribution of dryland regions (Budyko's dryness ratio)*

Extremely arid (D > 10)

Arid (D = 7 to 10)

Semi-arid (D = 2 to 7)

Sub-humid (D < 2)

Equatorial scale 1:88 000 000

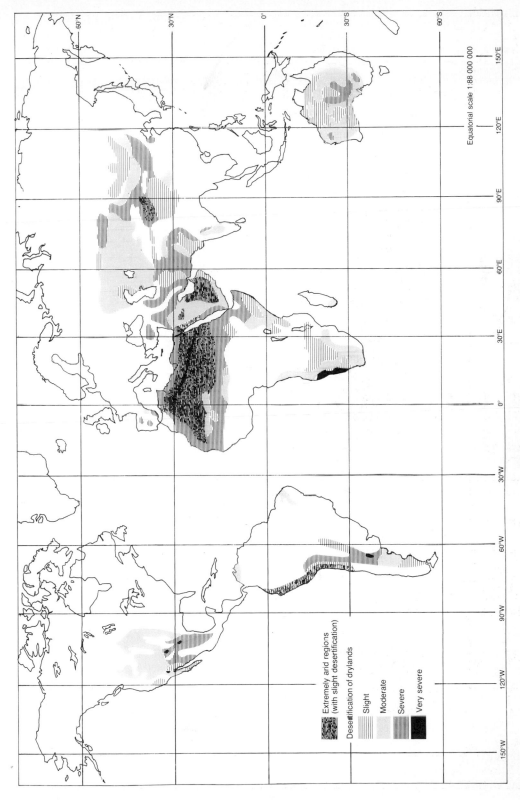

Figure 1.3 *Desertification of drylands*

Table 1.3 Criteria for estimating degree of desertification

Desertification class	Plant cover	Erosion	Salinization or water-logging (irrigated land) ECc × 10³ (mmhos)	Crop yields
Slight	Excellent to good range conditions class	None to slight	<4	Crop yields reduced by less than 10 per cent
Moderate	Fair range conditions class	Moderate sheet erosion, shallow gullies, few hummocks	4–8	Crop yields reduced by 10–50 per cent
Severe	Poor range conditions class	Severe sheet erosion, gullies common, occasional blow-out area	8–15	Crop yields reduced by 50–90 per cent
Very severe	Land essentially denuded of vegetation	Severely gullied or numerous blow-out areas	Thick salt on nearly impermeable soils	Crop yields reduced by more than 90 per cent

Note: ECc – electric conductivity, a measure of soil salinity

Table 1.3 indicates the criteria used to estimate the degree of desertification shown on the map. The importance of each criterion varies according to the land use in a particular area. Thus the major factor for grazing land is the destruction of vegetation cover; soil erosion is the major factor for rainfed cropland; while salinization (the build-up of salts) and waterlogging are the major factors leading to desertification of irrigated land.

Of secondary importance on grazing land are such factors as soil erosion, crusting (the development of a hardened surface) and loss of fertility, whereas fertility loss, crusting and compaction (the loss of soil structure) are of secondary importance on rainfed cropland. Compaction is also important on irrigated land where heavy machinery is used.

The world's extremely arid regions (Figure 1.3) are shown as being at slight risk from desertification. This is because the natural *productivity*, or growth of plants and animals, in such regions is extremely low and people have little or no impact on these ecosystems. However, for this very reason, some people argue that extremely arid regions are not at risk at all from desertification: they can hardly become more desert-like.

Desertification in the world's drylands as shown in Figure 1.3 is broken down by its level of severity in Table 1.4. About half of the global

Table 1.4 Desertification of the world's drylands

Desertification class	Land area (sq. km)	Per cent of drylands
Slight	24 520 000	52.1
Moderate	13 770 000	29.3
Severe	8 700 000	18.5
Very severe	73 000	0.1
Total	47 063 000	100.0

dryland area is slightly at risk. On the world scale about one fifth of the dryland area is severely desertified. The proportion varies between countries and continents. In Australia just 8 per cent of the country's arid area is severely desertified, while in North America the figure reaches 27 per cent, in South America 22 per cent, in Asia 20 per cent and in Africa 18 per cent.

When the extent of desertification is broken down according to major types of land use (Table 1.5) we see that grazing land and rainfed cropland are most severely affected, each being desertified on over three-quarters of its area. Irrigated land suffers less so, with about one fifth of the irrigated area in drylands affected by desertification.

Table 1.5 Extent of moderate to very severe desertification of agricultural land in dryland regions

Land use	Per cent desertified
Irrigation	21
Rainfed crops	77
Grazing	82

2.3 Regional difficulties

The global overview of the desertification problem shown in Figure 1.3 and Tables 1.3 to 1.5 appears to provide a neat summary of the areas suffering from this problem. A global summary of this kind is a useful first step to take when looking at such a large scale issue, but we must be aware that this summary is fraught with problems. Although a world view is useful from the point of view of large organizations such as UNEP, who have been charged by the United Nations with the responsibility of combating desertification, the real solutions to the problem can only be devised at a much more local scale.

When it comes to working out a plan of action to fight desertification at the regional or village level, more detailed information needs to be gathered on the way desertification is happening and its causes. Such information can only be obtained by long term monitoring of a particular area. For example, let us consider the range of possibilities that might explain a lack of vegetation cover on a village grazing ground.

Reasons might include:

1. too many cattle have been grazed there in the past and they have eaten all the grass;
2. the grass has been burnt to make way for planting crops;
3. the annual rains have failed;
4. the annual rains are late in coming;
5. the amount of water falling in the annual rains has been getting smaller for many years.

This is just one example of the sorts of problem that need sorting out before any effective control measures can be decided upon. We will return to the problems of actually measuring desertification in Section 4. One of the major difficulties of studying this subject is that there have been very few long term studies of desertification. This lack of information will be shown in the case studies of Part II.

Thus Figures 1.3 and Tables 1.3 to 1.5 should be treated with caution. Many scientists who work on the problems of desertification do not believe that it is yet possible to draw a world map of desertification because there are far too few local studies proving that desertification has happened or is happening. This is one of the key difficulties and controversies of the desertification issue. Some of the questions that we should be asking ourselves as we read this book include the following. Is desertification really happening at the scale that the world map in Figure 1.3 suggests? How should we define desertification? How can we successfully measure it?

3   THE CAUSES OF DESERTIFICATION

We have noted that the causes of desertification have been divided into two categories. Some scientists believe that human action is the culprit, while others believe natural factors to be responsible. Most workers occupy the middle ground between these two extremes, ready to investigate both human and natural forces in their quest to identify the causes of desertification. In this chapter we will take a detailed look at the various human and physical processes believed to be important in the promotion of desertification.

3.1 Human activities

The human activities which have been suggested as causes of desertification nearly always relate to the ways in which the land has been *overused*. In other words, they are methods of land use that have been employed too intensively, beyond the level at which the land can be used sustainably, leading to a degradation of the environment that may take years to recover.

There are many reasons to explain why certain land uses, some of which have been employed in drylands for centuries, become too intensive. For example, it may be that a method of cultivation or herding is seen to be too intensive in an area because that area is experiencing drought. A certain number of cattle can graze an area without causing long term damage (i.e. sustainably) during one decade, but during another, when drought occurs, the same number of cattle cause widespread damage to the few plants that survive the lack of rainfall. In this case we can see how difficult it often is to disentangle natural and human influences on the desertification process.

Increasing demand for food is probably the driving force behind overintensive land use in most world regions. Many developing countries have experienced rapid population growth in recent decades, with annual average rates of three per cent and more not uncommon. The population of many African countries has doubled in the last thirty years. More people means more mouths to feed. The extra food can either be produced from the same land, by *intensifying* production, or by agricultural *extensification*, using more land. Intensification may mean grazing more cattle on an area, or shortening fallow periods for fields. Extensification can mean using land that is not well-suited to agricultural use. Both processes have occurred in developing countries.

Demand for more food in developed countries in recent times has been met by the same methods. In the USA in the early 1970s, however, increased grain production was not a response to rising domestic demand, but the result of geopolitical reasons. The Great Plains farmers of the American South-West were encouraged by the government to increase grain production to sell to the USSR. This was to earn foreign exchange to help the national balance of payments deficit brought on by the high price of imported oil. Consequently, some 24 million hectares of idle cropland were brought back into production, extensifying production. At the same time, practices changed: crop rotations were abandoned, and farmers planted monocultures of corn, wheat and soya beans. New, more intensive technologies were also adopted, such as centre pivot irrigation (Figure 1.4), a system in which groundwater is used to irrigate great circles of crops two kilometres in diameter by a rotating sprinkler arm on wheels.

Figure 1.4 *Centre pivot irrigation, Great Plains, USA*

Although this activity brought prosperity to farmers, it was short lived as grain exports fell in the late 1970s and early 1980s. The impact on the environment of this phase was alarming. Land no longer protected by soil conservation measures adopted after previous ecological disasters, such as the *Dust Bowl* of the 1930s, was washed and blown away in large quantities. Government studies show that more than 40 per cent of US cropland is losing soil faster than it is being formed. Each year, some three billion tonnes of soil from US cropland is lost to wind and water erosion. The figure reaches five billion tonnes when erosion from pastures, grazing land and forest land is added.

The conclusion is simple, farming practices which were inappropriate to the local environmental conditions have resulted in desertification of large areas of the American Mid-West. Although this loss of soil should be reflected in declining crop yields from farmland, yields have been maintained by increasing the use of chemical fertilizers. But while fertilizers can maintain yields, they are costly and do not contribute to a good soil structure.

i) Overgrazing

Overgrazing is one of the main human activities which contributes to desertification. It results from the maintenance of herds whose numbers are too large for the land to support. An area of grazing land can support a certain number of animals without suffering any loss of quality, this is known as its *carrying capacity*. When the carrying capacity is exceeded overgrazing can have detrimental affects on vegetation, soil and eventually the health of the animals themselves.

Both the quantity and quality of vegetation can be altered by overgrazing. Soils become denuded and changes in the species mix can occur with the palatable grasses (those good for grazing animals) declining and giving way to invasions of drought-resistant species which are often less palatable and less able to bind soils together. The hooves of grazing animals also compact soil surfaces and break down the soil structure making the soil more easily eroded by wind and occasional rain storms. As the area of pasture and its quality declines, so the health of animals declines along with their production of milk and meat.

Overgrazing problems may be particularly acute around boreholes and wells where large numbers of livestock congregate (Figure 1.5). Other factors that have contributed to the

Figure 1.5 *There is clear evidence of overgrazing around this borehole in Kordofan province, Sudan (the borehole is on the right of the picture)*

widespread problems of overgrazing include the attitudes both of pastoralists and newly-formed developing world governments.

In the Sahel of Africa, for example, it is widely believed that good rainfall years in the 1950s encouraged pastoralists to expand their herds not only to increase production of foodstuffs but also because the size of a person's herd reflects their social status. Hence herds grew during times of good rain, but overgrazing was exposed by lower rainfall years during the 1970s and 1980s.

Government policies may contribute to overgrazing. The traditional methods that pastoralists have developed to live in the fragile dryland environment is to move their herds to pastures. True nomadic pastoralists move their animals almost continuously, whereas *transhumant* pastoralists follow set, often seasonal, routes to new grazing lands. While these methods are often the best, if not the only way of using such a changeable resource, the movement of peoples is often frowned upon by central governments who seek to control and tax their populations. Hence, in many cases there have been concerted efforts to settle, or *sedentarise* pastoralists, a policy that often results in localised overgrazing. However, the same problems can arise when nomadic or transhumant pastoralists move to wells and other water sources in retreat from drought conditions.

ii) Overcultivation

Overcultivation is another form of human action that can lead to desertification. It takes a number of forms. It may mean that pressure for more food encourages a farmer to cut the length of time a field is left fallow after cultivation. It may also mean the cultivation of soils that are not really suitable for the growing of crops.

Overcultivation reduces soil fertility and plant growth. Fields are often left fallow or used as pasture for several years after a period of continuous cropping. This is to allow the soil to recover its store of nutrients and rebuild its fertility and organic content that have been used by the growing crops. This method is often referred to as *shifting cultivation* in developing countries, the farmer moving on to plant another plot while the idle land is left to regenerate itself.

When a plot is not allowed to recover its fertility because fallow periods are shortened, the soil becomes exhausted, losing nutrients and suffering a breakdown of soil structure. Eventually, this has the effect of reducing crop yields and plant cover, which in turn leaves soils open to the ravages of water and wind erosion, so further depleting nutrient content and soil structure.

The erosional loss of nutrients from soils in drylands is usually greater than in more temperate climates. This is because nutrients in dryland soils tend to be concentrated in the upper layers since there is little rainfall to *leach* or wash them down through the soil profile. Thus, when soil is lost to blowing wind or flowing water it is the nutrient-rich upper layers which go first.

A similar pattern of events occurs when farmers are forced to cultivate land that is increasingly marginal for cultivation. This may happen simply through increased demand for food, but in many developing countries there is a stronger force behind this shift in cultivated area: the spread of cash crops for export.

Cash crops are those grown for sale in the city markets or for overseas export. Among the most widely grown cash crops in the Sahelian regions of Africa are groundnuts (peanuts) and cotton. In the developing world many governments encourage the growing of cash crops, often at the expense of subsistence crops. There may be a number of reasons for these priorities, three are particularly important.

1. Cash crops provide a major source of foreign exchange enabling developing countries to import goods and services they cannot provide for themselves.
2. Cash cropping is an agricultural system inherited from past colonial administrations.
3. Cash cropping is a farming practice that stems from industrialisation policies which often exclude the agricultural sector.

Whatever the motivation, the land area devoted to cash cropping in many developing countries has increased in recent decades. Niger, a former French colony in West Africa, provides an interesting example of the sort of chain of events that can lead to the problem of overcultivation. Niger's groundnut growing area doubled from 730 sq km in 1934 to 1420 sq km in 1954. By 1968 the area had increased again, by more than three times, to 4320 sq km. The increase in the 1960s was stimulated by private companies and the French government who were looking to encourage alternative sources of cooking oil in response to competition from US soya bean imports.

In the 1960s the farmers of Niger became heavily reliant on the income from groundnut sales. Food imports became necessary, but the terms of trade were becoming less favourable. At the same time the effects of groundnut cultivation on the soil were becoming evident. The crop so exhausts the soil that after three years of cultivation a plot should be left fallow for six years to recover.

To combat declining fertility the French government offered improved seeds and fertilizers, but these were hardly used because farmers were already in debt for seed and new technology. The circumstances of declining prices, declining soil fertility and mounting debt were made worse in 1965 when France was obliged to begin reducing its prices for groundnuts. Between 1967 and 1969 the price fell by 22 per cent.

To maintain reasonable living standards groundnut production was increased. This was partly done by using fields that lay fallow, leading to the overcultivation of large tracts of land. At the same time, traditional millet cultivation was pushed further into marginal areas, also resulting in overcultivation. The expansion of groundnut cultivation also affected livestock herders, pushing them on to marginal lands and depriving them of dry season grazing on formerly fallow fields. Thus overcultivation occurred in a number of ways. The result has been to promote desertification in Niger.

iii) Vegetation clearance

The clearance of vegetation by inhabitants of drylands is predominantly undertaken for two purposes: preparation of land for cultivation, and collection of wood for fuel.

Deforestation cannot really be divorced from the other problems of overcultivation and overgrazing. Increased pressures on arable lands often tends to push cultivation into forests and rangelands.

Wood collection for fuel is used for cooking and warmth in the cold desert fringe nights – daily temperature ranges of 20°C are common. Increasing demand from a growing population has produced a *fuelwood crisis* in many parts of the Third World. The crisis has not just stemmed from rural populations. It is perhaps more serious in the cities where urban inhabitants use more

Figure 1.6 *Collecting fuelwood is a time-consuming activity for women in Burkina Faso. Increasing population has led to faster vegetation clearance around settlements*

wood per head than their village counterparts. The consumption of fuelwood in Ouagadougou, capital of Burkina Faso, accounts for 95 per cent of national forest production in the country.

The effects of urban fuelwood collection are strikingly evident in the almost circular areas of treeless land that surround major cities in dryland regions. In 1955 dense acacia woods were common around the fringes of Sudan's capital Khartoum, but today only isolated pockets of woodland survive within 100 km of the city. Almost all the trees within a 40 km radius of Ouagadougou have been felled (Figure 1.6).

The effects of vegetation clearance are similar to those of overcultivation and overgrazing: a degraded vegetation cover exposes soils to erosion and breakdown of soil structure. Trees also give shade to people and animals and are a source of food for livestock. Their removal also affects the water table. Furthermore, as trees are removed from surrounding areas, villagers supplement wood for fires with dried animal dung. This dung would otherwise be left on the soil, acting as fertilizer. Collection and burning of dung, therefore, adds to decreasing soil fertility.

iv) Salinization and waterlogging

Salinization, the build up of salts in soils, is a common problem faced by irrigation schemes in dryland areas. Irrigation of arid soils should be the answer to dryland food production, ironing out the vagaries of unreliable rainfall. Unfortunately, many irrigation schemes record declines in yields after just a few years of operation. In the extreme, salinization occurs to the point where the land is too salty to sustain plant life.

The key to successful irrigation is good drainage. All too often irrigated land is not drained properly, the groundwater in the soil begins to rise and as it does so, evaporation by the hot desert sun increases the concentration of its salts. On reaching plant roots, the salty groundwater impairs plant growth. Sometimes soils become completely waterlogged and salt crusts form on the surface. If the problems are not corrected the land eventually must be abandoned.

According to some estimates, the amount of land being brought under new irrigation works every year is balanced by the amount of productive land lost to salinization and waterlogging. In Pakistan, two-thirds of the country's

150000 sq km of irrigated land suffers from salinization, waterlogging or both. Each year an estimated 40000 hectares are lost to these problems. In Egypt, no less than 90 per cent of irrigated land is affected by waterlogging and 35 per cent suffers from salinity problems. In the USA up to a quarter of all irrigated land is affected by salinization (Table 1.6).

Table 1.6 Salinization of irrigated land in selected countries

Country	Area of cultivated land under irrigation (million hectares)	Percentage of irrigated land affected by salinization
Australia	1.6	15–20
China	45.4	15
Egypt	2.5	30–40
Iraq	1.8	50
Pakistan	15.3	35
Spain	3.1	10–15
USA	19.8	20–25

3.2 Natural factors

i) Drought

The most obvious natural cause of desertification is a lack of rainfall, or drought. A drought can be thought of as the continued absence of expected rain. The effect of drought on plants is to cause wilting and, if prolonged, death. For human occupants of dry, drought-hit areas the effects on cultivation and livestock are severe. Clearly, therefore, drought can lead to desertification.

One aspect of droughts that should be made clear is that they are not abnormal phenomena. Although they often cause disruption to human society, plants and animals, droughts are a perfectly normal aspect of the climate of many world regions, and particularly so in drylands where rainfall is always sporadic and unreliable.

Meteorological drought is the term used for a drought which uses simple rainfall data to describe it. Because of differences in the variability of expected rainfall between the world's climate regions, the definition of meteorological drought varies across the world's climatic zones. In the British Isles, for example, a so-called *absolute drought* is a period of at least 15 consecutive days during which no more than

0.2 mm of rain is reported on any day. In the world's drylands, where less rain is expected, a period of several months without rain may occur before a drought is officially declared. As a general rule, the larger the amount of rain expected, the shorter is the period of deficiency to be termed a drought. However, in extremely arid desert areas, where minimal rain is expected, the term drought is not of much use. In the same way that aridity is not always best defined in terms of rainfall alone, so drought can also be defined in more complex terms.

Agricultural drought is a more involved and, for some studies, more appropriate form of drought. It is defined by the US Weather Bureau as:

> 'A period of dry weather of sufficient length and severity to cause at least partial crop failure.'

The essence of agricultural drought relies on the calculation of what is termed *effective rainfall* which can be thought of as the amount of rain that is actually available for plant growth. A rain storm brings a certain quantity of water to a portion of land, but some part is often lost by such means as evaporation or by flowing off the ground surface.

A *water balance model* is used to calculate effective rainfall. It is an equation which can be worked out for a certain area and expresses the water remaining when the input of precipitation (P) is reduced by outputs of evapotranspiration (Et) and changes in surface runoff, drainage and soil moisture storage (S). Hence:

$$\text{Effective rainfall} = P - Et +/- S$$

Physiological drought is a third form of drought that is relevant to desertification studies. This is a term used to describe the condition of plants that suffer from an excess of salts, usually in poorly-drained irrigated areas. In such circumstances it is not so much a lack of water that is harmful to plants, rather the high concentration of salts means that plants cannot properly use what water is available.

In all cases drought has the same result: plants die and vegetation cover is reduced. Thus drought can lead to desertification. But in most dryland areas, plants are well-adapted to drought conditions so that although their foliage may be lost during drought periods, their seeds are dispersed ready to spring into life again when rain returns. For the scientist who seeks to determine whether a decrease in vegetation will continue beyond the duration of drought, this property of dryland plants means that studies must continue throughout and beyond drought periods. Only in this way can we determine whether the loss of vegetation is due solely to lack of rainfall or whether human factors such as overgrazing are also partly responsible.

ii) Causes of drought

The causes of meteorological and agricultural droughts in the world's drylands is the subject of much debate. Rain usually falls from an *unstable* atmosphere. This means that the air is not still, but subject to vertical movements. These vertical movements are necessary for clouds and rain to form because as air rises it usually cools, and what moisture it contains starts to condense out into liquid.

Most rain that falls in mid-latitudes is formed in large meteorological systems known as *depressions* or *cyclones* in which rainfall may occur over hundreds of kilometres. In drylands rain tends to form and fall from much smaller systems, such as thunderstorms, which precipitate over smaller areas measured in tens of kilometres. Since a drought means little or no rain, it is encouraged by generally *stable* air which does not allow such rain-bearing systems to form. The world's drylands occur in areas where the general circulation of the atmosphere maintains stable conditions most of the year, but the reasons for the persistence of stable air masses in certain places are not all clear.

The cause of drought in the Sahel region of Africa (see Figure 2.1), for example, which has continued more-or-less unabated since the late 1960s, is still uncertain. Some of the most recent studies in this region have found strong relationships between rainfall in the Sahel and the temperature of the surface waters of the tropical parts of the Atlantic Ocean. It appears that years of deficient rainfall in the Sahel are linked to years when the equatorial Atlantic surface waters are cooler than usual. The link could be that a cooler-than-normal sea surface releases less moisture into the air so that less is available over the Sahel to form rain. But just why the sea surface temperature of the Atlantic varies in this way is still to be resolved.

Similar global links between sea surface temperature and droughts have been identified from the Indian and Pacific Oceans. In the case of the Pacific, a large area of equatorial surface water off the coast of South America is occasionally much warmer than normal around Christmas time. The timing of these events has meant that the name *El Niño* ('the little child') has been given for such occurrences. This change in sea surface temperature, which can be as much as 10°C warmer than normal, appears to trigger abnormal rainfall in many parts of the world, with severe droughts being recorded as far apart as the Sahel and Australia (see Dawson, 1991, Global Climatic Change).

iii) Climatic change

Droughts are normal events in the climate of dryland areas. They have occurred repeatedly throughout the periods for which meteorologists have records and are certain to occur in the future. But the time between consecutive droughts and the duration and intensity of a drought is often very difficult to predict.

Figure 1.7 shows rainfall data for the meteorological station at Dori in the Sahel region of West Africa. This graph shows several important aspects of Sahelian rainfall. First, there are marked differences in totals from year to year. In 1969 for example, rainfall was more than 250 mm greater than in 1968. Secondly, there have been several long periods of above or below average rainfall. A drought struck across the whole Sahel from 1910 to the early 1920s – the tail end of this drought is indicated in Dori's rainfall figures. The 1950s was a decade of above average rainfall. Since the late 1960s there has been an extended drought period.

The duration of the most recent drought in the Sahel has tempted some scientists to suggest that the area is experiencing a significant change in its climate. Others have disputed this claim. They suggest that the drought of the 1970s and 1980s is not abnormally long for this type of dryland climate.

The question of climate change is very important for desertification studies. If the climate of dryland areas, such as the Sahel, is changing to become drier, this has important implications for the future of desertified areas.

But although scientists have very good evidence to suggest that many world regions have experienced marked changes in their climates during geological time, recent climate change has yet to be proven in the Sahel or any other dryland

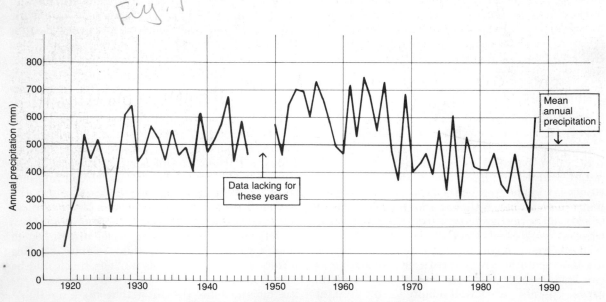

Figure 1.7 *Rainfall at Dori, Burkina Faso*

area. The great degree of interannual variation in dryland rainfall totals makes the question of climate change very difficult to answer. These great fluctuations make it difficult to identify any trends that may be happening.

In the near future, however, these questions will become increasingly important. The effect of global warming due to the *greenhouse effect* may soon make an identifiable mark on the rainfall of the Sahel and other dryland areas.

Figure 1.8 *Before and after rainfall in a dryland area of the USA. For a short time after a downpour the desert suddenly becomes green. Plants such as these Saguaro cacti respond quickly to an input of moisture, and the one on the right is in flower. These pictures illustrate the drylands' adaptation to sporadic rainfall. For scientists studying desertification this dynamism complicates the questions of whether permanent change is occurring*

4   THE CONSEQUENCES OF DESERTIFICATION

Many of the consequences of desertification have been hinted at in previous chapters. Those for the physical environment include increased erosion and breakdown of soil structure, loss of vegetation cover, changes in the types of vegetation species, increased soil salinity and the lack of surface and available water. For the human population of desertified areas the consequences include loss of livestock, loss of crops and loss of fuelwood, which result in general shortages of food and fuel. These circumstances, when they occur in the extreme, can ultimately lead to economic impoverishment, famine and mass migrations of *ecological refugees* from desertified areas.

In this Section we shall look first in more detail at the consequences for the human populations of desertified areas. Since these human and physical consequences are the products of desertification, we will also examine them in more detail to assess their potential use as indicators of the process. But before proceeding to look at possible indicators, there is another set of consequences that will be discussed. These are consequences that some scientists believe may be acting to prolong the agony of desertification. They can be referred to as *feedback mechanisms*.

4.1 Effects on society

The impact of desertification on human society is very closely related to the effects of drought. The dramatic years of the *Dust Bowl* in the Mid-West of the USA and the recent years of Sahel drought serve as good illustrations of the way in which prolonged drought and desertification can have long-lasting effects on people's lives.

i) Migration

The Dust Bowl saw unprecedented soil erosion in massive dust storms during the drought years of the 1930s in the Great Plains of the USA. From the late 1870s the grasslands of the Great Plains had been slowly transformed into a wheat-growing belt. Waves of new settlers arrived in the area between 1914 and 1930, at the same time as mechanised agriculture was widely adopted. When drought hit the area in 1931 the newly-exposed ploughed soils were blown away on a massive scale. One event, that blew soil from Montana and Wyoming on 9 May 1934, carried

Figure 1.9 *This famous image, taken in Oklahoma in the 1930s, captures the plight of farming families in the Dust Bowl*

350 million tonnes of dirt eastwards. In the morning of 11 May this dust was settling on the streets and cars of Boston and New York.

The decade of the *Dirty Thirties*, as they have become known, saw permanent changes in the lives of hundreds of thousands of Americans. The Dust Bowl came at the same time as the *Great Depression* in the USA, and many farming families lost all they had in the dust. The result was a large scale migration of farmers and their families away from the Great Plains states of Texas, Kansas, Oklahoma and New Mexico.

Almost one million people left their farms on the Great Plains in the first half of the 1930s. After 1935 another 2.5 million walked out of their wooden houses and took to the road, carrying with them as many of their possessions as they could. Many people moved into nearby towns and looked for alternative work. Many others moved much further, towards the west.

Between 1935 and 1940, the state of California was swamped by 300 000 poor people, ecological refugees from the Dust Bowl. Most came from Oklahoma. They were called 'Okies' and they were often treated like third class citizens. The plight of these people inspired a number of writers and artists of the times. John Steinbeck's novel *The Grapes of Wrath*, the story of one family's flight from the Dust Bowl, has become one of America's classic novels. The poems and songs of Woody Guthrie also record the social upheavals of the Dirty Thirties.

ii) Traditional responses and preparation

The movement of people away from zones of disaster is a common response to such ecological catastrophes as drought and desertification, often leading to mass migrations from affected areas. During the drought years of the 1970s in the Sahel, for example, many herders moved into cities after seeing their livestock die through a lack of pasture.

But unlike many of the new farmers of the US Great Plains, long term inhabitants of many dryland regions have grown accustomed to their natural problems. In the villages of northern Nigeria there was a range of traditional responses by farmers whose millet yields were cut by more than half and whose cattle, sheep and goats died in large numbers during the 1970s. The most common response from small farmers

was to work on the land of larger farmers, thus earning money to buy food they were unable to produce themselves. At the same time many families turned to *famine foods* which are traditionally used in times of drought and food shortages. These famine foods are mostly leaves and are only occasionally eaten in normal times. Other approaches to dealing with the effects of drought include making mats and ropes to sell to traders in the markets and chopping wood to sell for fuel. The social structure is also important during times of crisis: a complex network of extended families can help poorer relations with gifts of food during bad times.

These local ways of coping with the situation are supplemented by strategies 'away from home'. Adult males often take to the road in northern Nigeria, to look for work in the south or in the cities, but during famine years this movement increases until whole villages can become almost deserted.

In addition to these traditional responses to drought when it hits, this area of northern Nigeria is typical of many semi-arid zones in that local people have ways of preparing for drought which they know will always return. This traditional insurance has four main approaches in northern Nigeria.

1. Produce and store food during good years, the oldest insurance system in history.
2. Work harder at alternative activities to save money to pay for food when necessary.
3. Rely on the community network of extended families. This network is supplemented by relatively rich members of society who have a moral responsibility under Islam to help poorer neighbours in hard times.
4. The traditional insurance of diversified farming and herding: growing several different types of crop; having several fields in different places; keeping different sorts of livestock such as cattle, goats and sheep; and moving animals to where pasture is growing. The combination of these methods gives some protection against hazards which are particular to a specific crop, animal or place.

One of the main reasons behind the large scale of human suffering in the Sahel during the 1970s and 1980s is that these traditional ways of coping

Figure 1.10 *Nouakchott, capital of Mauritania in West Africa, has grown very rapidly as a consequence of twenty years of drought and desertification. Migration of desert nomads to the city has swelled its population from 20 000 in 1960 to more than 350 000 by the end of the 1980s. Most of these ecological refugees live in unplanned makeshift housing like that shown in the foreground of this picture.*

with the problems of the environment are under threat.

1. The possibilities for grain storage have become more limited by a lack of land and the increase in the number of mouths to feed.
2. The opportunities for paid employment have in some areas declined. In Nigeria, for example, herders from Niger who had settled in northern cities were rounded up and sent home in the early 1980s. Many of the *urban informal service sector* enterprises such as traders' tables and kiosks were demolished and their self-employed owners sent back to the land.
3. The drought and population growth of the 1970s meant that many of the people who might have been relied upon to help their poorer relations did not have enough food for themselves.
4. Most agricultural development projects in the Sahelian countries have encouraged farmers to specialise in particular crops or animals, so eroding the traditional insurance of a diverse farming approach.

iii) Famine

People starving to death is an image that many of us associate with the drought-prone regions of Africa. Famine, an extreme shortage of food over a long period, can certainly be linked to the effects of drought and desertification. Lack of rainfall and decline in biological productivity of an area will lead to less food growing and thinner animals to milk or eat.

But famine is a complex situation which is often affected by many other factors besides drought and declines in productivity. *Population growth* is a key factor in the famine equation. In Ethiopia and Sudan, for example, two countries where terrible famines have hit during the 1980s, the rate of population growth has been close to three per cent a year during that decade. In Ethiopia this has meant an extra 1.3 million mouths to feed each year. In Sudan the figure is more than 600 000.

Although food production has risen steadily over the last twenty years in both countries, it has hardly kept pace with the rapid rise in population. Thus in Ethiopia the food produced *per person* has fallen, while in Sudan it has increased only slightly.

War and civil strife is another important influence on famine. Both Ethiopia and Sudan experienced prolonged civil wars during the 1980s, which disrupts societies and economies. This disruption also contributes to the *poverty* of many individuals which is another major factor in the occurrence of famine, simply because people who have no money cannot afford to pay for food. In the words of one researcher into the causes of famine: 'A general shortage of food is hardly ever the cause of famine. Famines are caused by the poverty of victims: they are unable to purchase the food that is available.'

Thus, although famine may be popularly associated with drought and desertification, it is important to remember that these are usually relatively minor influences on the situation which leaves large numbers of people without food.

4.2 Feedback mechanisms

Two physical consequences of desertification may act to prolong drought conditions and desertification.

i) Albedo change

One suggested mechanism for prolonging drought results from the loss of vegetation from a ground surface or its *denudation*. This comes about due to the changed nature of the ground surface. All surfaces, when they receive solar radiation, absorb part of this radiation and reflect the remainder back into the atmosphere. The amount reflected depends on the properties of the surface, most importantly its colour and texture. The reflectiveness of a surface is known as its *albedo*. When a ground surface suffers from a loss of vegetation its albedo changes. In drylands the albedo increases, so that more incoming solar radiation is reflected, because a bare ground surface is a lighter colour and has a generally smoother surface texture than a vegetated one.

The proposed feedback mechanism that results from this increase in surface albedo works in the following way. Because more radiation is lost by reflection, the ground surface would be cooler. This would mean that the lowest levels of air, which are warmed by contact with the ground, would also be cooler. The result of this cooler lower air layer would be that air above tends to sink downwards. Since rising air is needed for cloud and rain formation, the outcome is less rainfall. The feedback would then continue over a long period because less rain would further reduce vegetation and so the process would feed off itself.

ii) Atmospheric dust

The other feedback idea stems from the large amounts of soil dust that can be blown from denuded soil surfaces. Dust lifted from dry, bare terrain by strong winds fills the air, often to heights of several kilometres. This atmospheric dust may also act to suppress rainfall. This is because dust-filled air layers tend to be stable, and instability is necessary for convective rain cells to form.

Again, the theory is a reasonable one, and large increases in dust storms have been recorded in desertified areas such as the Sahel, but the theory is still to be proven.

4.3 Monitoring the consequences of desertification

The range of desertification consequences outlined at the start of this section should make a reasonable set of indicators that could be used to measure and monitor the process. The map and information on the global extent of desertification in Section 2 are based on expert assessments. But they are no more than estimates. The true situation is not well-known, simply because there have been remarkably few studies to measure or monitor desertification *in action*.

Some of the studies and evidence for desert-ification that does exist is reviewed in Part II, but in this section let us take a brief look at the methods that could be used to monitor desert-ification and some of their problems.

Some suggested indicators are shown in Table 1.7. Some factors are more directly indicative of desertification than others. Many changes in human well-being, for example, may reflect outside social and economic influences. Other changes, such as to livestock herds, can be indirect, or unrelated. Are large herds, for example, telling us that pastures are plentiful or

that overgrazing is going on and therefore that future desertification is on the way?

Table 1.7 Possible indicators of desertification

Increase in soil salinity levels
Increase in dust storm frequency
Increase in surface albedo
Increase in soil compaction
Decrease in biological productivity
Decrease in biomass
Decrease in soil organic matter
Sand dune mobilization
Loss of key plant species
Loss of key animal species
Domestic livestock herds: loss of production/yield
Decreased human well-being

Some of the indicators, such as a decline in biomass, an increase in dust storm frequency and higher levels of soil salinity, can be direct reflections of desertification. But care must be taken to establish just by how much the biomass, for example, must decline before we allow ourselves to state that desertification is happening. Do we need to monitor continuing decline over a certain period? Or does a dip below a certain level or *threshold* indicate the onset of desertification? Only long term research can provide us with answers.

Another important aspect of desertification monitoring is the scale of the study. If monitoring is for a village and surrounding region, different indicators may need to be chosen to those suitable for a national, continental or global study. In the latter case, the most appropriate indicators are those that can be monitored from an aeroplane or satellite. But at the local level the quality of such data will certainly have to be supplemented with field studies.

An even more crucial aspect of the study is the time scale adopted. How does the scientist determine a significant or permanent change from the *normal* situation? Drylands typically experience sporadic or seasonal rainfall and droughts, and the vegetation therefore typically changes in its coverage (see Figure 1.8), both sea-sonally and from year to year. Monitoring, there-fore, must take place over several years and great care must be taken over choosing the appropriate periods at which to take measurements.

In this part of the book we shall look at some of the areas in the world where desertification is supposed to be in progress. We shall examine the evidence for desertification, the consequences of the process and some of the actions that have been suggested and taken to combat the problem.

1 DESERTIFICATION WITHIN THE SAHEL

The Sahel, a desert-marginal zone that stretches across Africa, is the area that first brought desertification to the world's attention. Sahel is an Arabic word meaning *shore*, which has been given to the strip of land along the southern border of Africa's Sahara Desert. It stretches 5000 km from Senegal and Mauritania on the Atlantic coast to Chad and portions of Sudan and Ethiopia in the east (Figure 2.1). The area covers about 20 per cent of Africa's 30.3 million square kilometre land surface.

On a vegetation map of Africa, this zone is

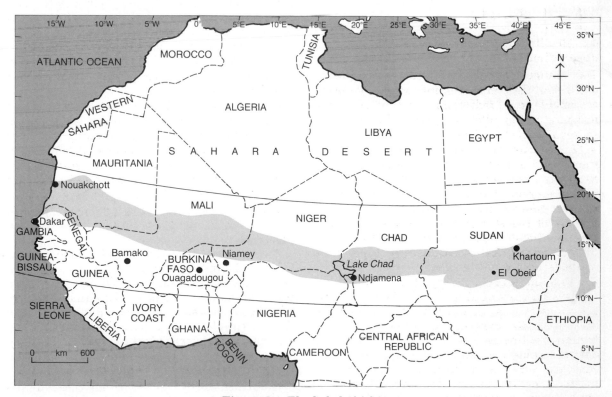

Figure 2.1 *The Sahel of Africa*

marked as savannah grasslands, a transition area between the arid desert to the north and the more humid woodlands to the south. On a map of African annual average rainfall the zone can be taken as that lying between the 100 mm and 600 mm isohyets. But although we can neatly draw lines on a map to delimit this area we must remember these lines can only give us a general indication of the extent of this zone. As we have seen in Part I, one of the key characteristics of these semi-arid zones is their natural changeability. It is this normal changeability that makes our efforts to identify and measure the process of desertification so difficult.

The desertification problems of the Sahel will be looked at in two countries where the best data and information are available: Mauritania in the west and Sudan in the east.

1.1 Mauritania

The Sahelian zone of Mauritania (Figure 2.2) is an area where desertification has been occurring for the past twenty years. But the problem of whether human action or natural factors are to

---

Mauritania in West Africa covers over one million square kilometres, almost the area of France and Spain combined. Much of this area is in the Sahara Desert, with a strip of Sahelian grassland in the southern half of the country dividing it from the fertile plains of the River Senegal which forms the border with neighbouring Senegal.

Mauritania's population is small, two million in 1987. Traditionally, most of these people are nomadic or partly nomadic herders of goats, sheep, cattle or camels. There is some irrigated agriculture on the River Senegal where millet, dates, rice and sorghum are grown. Otherwise, the mainstay of the country's economy is the mining and export of iron ore from reserves in the northern desert and the coastal fishing industry. The country is one of the world's poorest, with an annual income per person of about US$400. From 1920 it was a separate colony of French West Africa until 1960, when Mauritania gained independence. Before Mauritania had celebrated its first ten years of independence, the country was caught in the grip of a prolonged drought that began in the late 1960s and has continued throughout the 1970s and 1980s.

Figure 2.2 Mauritania in context

---

blame is a crucial one in this example. The period for which the evidence for desertification is available is during a period of intense drought. Thus, we must first understand the climate of the region.

i) Climate

The surface wind circulation and the seasonality of rainfall in this part of the western Sahara and Sahel is very largely determined by interactions between two high pressure systems and the *Intertropical Convergence Zone* (ITCZ) (see Figure 2.3).

The pressure systems are known as the *Azores High* and the *Saharan High*. The ITCZ is a frontal zone which marks the meeting place between warm, moist air from the Gulf of Guinea to the south and hot, dry Saharan air to the north. This ITCZ moves north and south with the seasons, bringing West Africa's seasonal rainfall when it comes under the influence of the moist air from the south.

In the winter months the Azores High is close to the African coastline and the Saharan High is strong, while the ITCZ is at its southernmost position over the Gulf of Guinea (Figure 2.3A). The circulation from the two pressure systems creates a windflow that is from the north-east or east over much of the western Sahara. These are usually hot, dry winds from the desert, with little moisture. During the first six months of the year these hot, dry winds generate fierce dust storms in Mauritania, blowing loose soil, sand and clay particles from the dry, unvegetated land surfaces.

In the summer months, the Saharan High is replaced by a shallow low pressure system in the central Sahara and the Azores High is strong and further from the coast (Figure 2.3B), giving more northerly winds at most stations. During the period from June to September the ITCZ is also at its most northerly position, so that the Sahelian areas of southern Mauritania receive south-westerly monsoon winds that bring the seasonal rainfall to the area. The northward progression of the ITCZ, being the boundary of the moist air from the Gulf of Guinea, is reflected in the way that the most southerly parts of Mauritania receive greater annual average rainfalls, beginning slightly earlier in the year, than those parts further north, as shown in the rainfall diagrams in Figure 2.4.

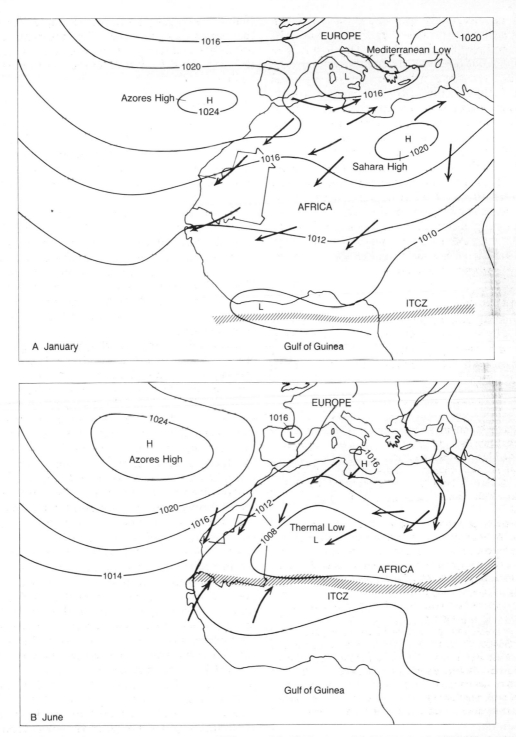

Figure 2.3 *Sea level pressure contours, in millibars, of the Sahara and adjacent regions in January and June. Average seasonal trends in surface wind flow are indicated by arrows*

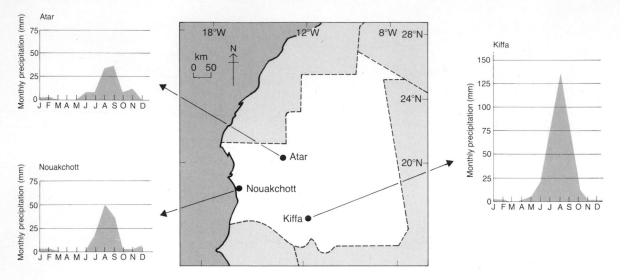

Figure 2.4 *Monthly rainfall totals in Mauritania*

ii) Increasing dust storms

Since the onset of drought conditions in Mauritania in the late 1960s, the occurrence of dust storms has increased. Dust storms are a measure of soil erosion and thus may be indicative of desertification.

The increase has been documented using data from meteorological observing stations. These stations report dust storms when the observer's visibility at eye level is reduced to 1000 metres or below. The observer takes the visibility readings using a number of known objects at set distances from the observing station. The method is used by meteorological observers throughout the world, but because the observations are taken by individuals some variation is likely.

There are other problems with these types of data. For example, it may be difficult for the observer to distinguish between dust in the atmosphere and fog or pollution. Another common problem, particularly in developing countries, is that many stations are not manned 24 hours a day and therefore events such as dust storms, which may last less than an hour, may not always be recorded. Nevertheless, despite the problems with dust storm data, they represent a reasonable method for monitoring changes in the environment.

The variations in the annual number of dust storm days (a day on which a dust storm is recorded) and annual rainfall totals at Nouakchott are shown in Figure 2.5. The obvious trends in the two lines on the graph are quite apparent: declining rainfall totals are mirrored by increasing dust storm frequency.

The main onset of the drought can be seen in the rainfall totals for 1970 and 1971 when just 48.1 mm and 17.9 mm fell. These totals represented just 32 per cent and 12 per cent of the long term average for the station over the period 1949–67. The number of dust storm days increased dramatically from six in 1970 to 65 in 1974 before a reasonably high rainfall year in 1975. Dust storm activity then declined in 1976 and 1977.

In 1977 the rainy season brought only 2.7 mm of precipitation making it the driest year since records began in 1931. Dust storm activity again rose to new heights in 1978 and 1979. After a relatively heavy annual rainfall in 1979, the number of dust storm days dropped to 33 in 1980, but rose again to an unprecedented 83 days in 1983 as the rains failed again and remained at this high level until 1986.

Detailed analysis of dust storm data for seven other stations in central and southern Mauritania shows that on average, dust storms had increased by more than six times across the country during the first ten years of the drought.

The effect of drought on dust storm occurrence

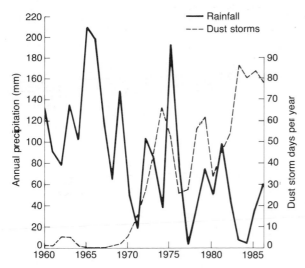

Figure 2.5 *Annual variation in rainfall and dust storm days at Nouakchott, Mauritania*

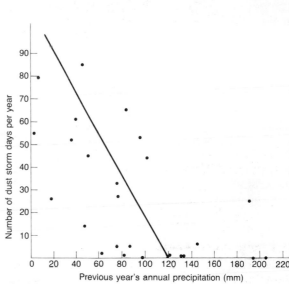

Figure 2.6 *Frequency of dust storms in relation to previous year's rainfall at Nouakchott, Mauritania*

at Nouakchott is further investigated in Figures 2.6 and 2.7. Figure 2.6 compares dust storm frequency to the previous year's rainfall because, as described above, dust storms usually occur in the first six months of the year, before the onset of the rainy season. A general trend can be seen in these data: declining rainfall amounts in the previous year are related to increasing numbers of dust storm days. A simple correlation test of these data, however, shows that the relationship is not very significant. In other words, there is an unreasonably high likelihood of this relationship occurring by chance. This means that we cannot be sure that a previous year's rainfall has any effect on soil erosion by wind.

Figure 2.7 goes a little further than Figure 2.6 to compare a year's dust storm frequency with the average rainfall of the previous three years. Here the relationship is stronger and statistically significant. What this graph tells us is that poor rainfall over a number of years has an effect on dust-raising. This can be explained by the cumulative effect of low rainfall on vegetation and soils: vegetation dies off and soils dry out and crumble to form smaller, less consolidated particles. This leaves the soil less protected and more easily blown away by the wind.

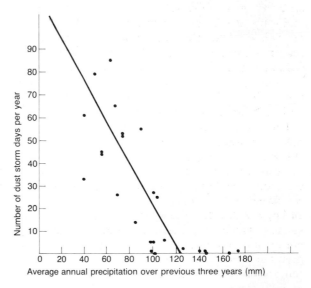

Figure 2.7 *Frequency of dust storms in relation to mean annual rainfall over the previous three years at Nouakchott, Mauritania*

iii) Natural or human action?

It is clear that dust storm activity has increased significantly in the Sahelian latitudes of Mauritania during a period of intense drought. The graphs in Figures 2.6 and 2.7 confirm that a lack of rainfall plays some part in explaining this increase in soil erosion by wind.

But the effects of reduced rainfall on the *erodibility* of a soil (the ease with which it is eroded) can also be brought about by human actions.

There are relatively few good studies of desertification in Mauritania, but reports from a number of authorities suggest that there was a widespread overuse of the environment both during the drought years themselves and before the drought began.

A report for the United Nations Environment Programme in 1985 suggests that environmental problems in Mauritania were due to overgrazing and the excessive cutting of woody vegetation especially for fuelwood. These activities were causing problems before the drought hit, but the prevailing dry conditions of the 1970s had the effect of amplifying them. The drought also encouraged herdsmen to cut woody vegetation to provide fodder and later forced them to concentrate in the southern parts of the country and Senegal. Sand encroachment and shifting dunes also affected many areas, in particular around urban centres and settlements.

Some authorities have produced estimates of the number of head of cattle in the country before and during the drought years. Between 1959 and 1968, which were generally good rainfall years, cattle numbers increased from 1.25 million head to 2.3 million. However, overgrazing of pastures and the desiccating effect of the drought saw cattle numbers fall again to near one million by 1973. It seems likely, therefore, that human action also contributed to the increased soil erosion by wind shown in Figure 2.5.

iv) Human consequences

Whether the widespread loss of vegetation and increased soil erosion by wind in Mauritania during the 1970s and 1980s was the result of drought alone or drought and human action, the consequences for Mauritania's people have been dramatic.

Twenty years ago, one out of two Mauritanian children was born in the desert, but today nomadism is vanishing. People have migrated to the south of the country and the cities, where they can live on shipments of food aid. In 1965 less than half of the country's citizens lived in cities, but by 1986 this figure had reached 85 per cent, their numbers swelled by the ecological refugees from drought and desertification.

In 1960, Nouakchott was a small town of 20 000 people. By 1988 the number had reached 350 000, more than half of whom are refugees. In the refugee slums around the city (page 24), one in three youngsters do not have enough to eat and the average life expectancy is just 46 years.

v) Solutions

Whether the consequences of desertification in Mauritania will be resolved should rains return is an unanswered question. One solution that has been attempted to reduce the effects of wind erosion around Nouakchott was to plant a *green* belt of trees around the town. This project was started in 1975. Figure 2.5 suggests that this method was inappropriate and has not been successful in preventing the blowing of dust storms.

vi) Summary of the evidence

Mauritania provides an example where we have a number of different types of evidence for desertification. We have scientific evidence for increased soil erosion in the form of numbers of dust storms. A widespread loss of vegetation has also been documented in personal observations by field workers, although there have been no systematic surveys of vegetation cover before and during the drought. Nevertheless, these lines of evidence suggest that desertification has been occurring in the Sahelian parts of Mauritania.

The causes of desertification are probably both climatic and human. But while we have proved statistically that a lack of rainfall has had an influence upon dust storm frequencies we are not able to make a similar statement about the effects of human action.

1.2 Sudan

i) Measurement of vegetation shift

Sudan (Figure 2.8) was the scene of the first and only real study which set out to measure the shift

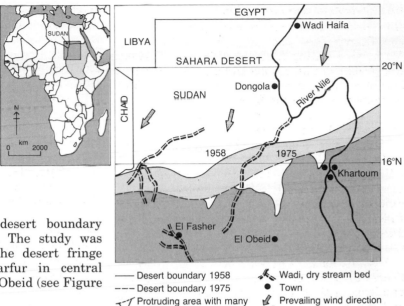

in vegetation zones at the desert boundary resulting from desertification. The study was carried out by Lamprey in the desert fringe regions of Kordofan and Darfur in central western Sudan just north of El Obeid (see Figure 2.1).

Sudan is Africa's largest country with a land area of some 2.5 million square kilometres. The country's climate varies from the arid north, part of the eastern portion of the Sahara Desert where many of the inhabitants are nomadic, through savannah grasslands of the central part, to hot, wet tropical lands in the south. Sudan's population is about 22.4 million. The country gained independence from Britain in 1956. Agriculture is the mainstay of Sudan's economy, with many crops for export grown in irrigated plantations on the River Nile in the east. Sudan is one of the world's poorest countries, with average income per head of around US$400.

Figure 2.8 *Sudan in context*

This report is vital to our understanding of the way in which desertification is presently viewed. It has become one of the most widely cited references on the rate of spread of the Sahara and is the basis for most statements about the rate of worldwide desert advance. It is therefore important to spend some time looking at the methods used for the study.

Lamprey's study was carried out in 1975. His team travelled over the desert fringe area in a light aeroplane and a ground support vehicle, identifying the position of the boundary between grassland and sub-desert scrub and the true

Figure 2.9 *Advancing desert boundary as found by Lamprey's study in Sudan. The conclusion was that the Sahara had advanced 90 to 100 km between 1958 and 1975, an average rate of about 5.5 km a year, but see text for criticisms of this study (also see Part I)*

desert. Although this boundary was diffuse, it was deemed possible to map it to the nearest 5 km. The approximate desert boundary that they identified was then compared to the boundary mapped in 1958 by a team who were zoning Sudan's vegetation.

The two desert boundaries, for 1958 and 1975, are shown in Figure 2.9. Also shown on this map is a vast area of moving desert dunes that had not been mapped in the 1958 study, but were, in 1975, apparently drifting southwards, engulfing the scrub vegetation and threatening fields and villages.

The findings of the Lamprey survey have been questioned, in particular by a team from the University of Lund, Sweden. Their work on the same part of Kordofan was completed in the 1980s using satellite imagery from the Landsat satellites and aerial photographs taken during the period from 1961 to 1979.

The Swedish team found no evidence to back up the claim that the desert margin had been moving southwards. Among other things they could not find the extensive area of sand dune encroachment that Lamprey had mapped. Also the distribution of cultivated land in 1962 was about the same as in 1979 and there was no systematic change in the size of degraded land areas around settlements or water sources.

The Swedish team also looked at the methodology behind the drawing of the 1958 vegetation boundary that Lamprey had taken as a base line to conclude that the desert had advanced. They suggested that the boundary used was simply drawn at the 75 mm isohyet, which is 90 to 100 km north of the 100 mm isohyet, rather than actually mapped by survey. Their estimate of the desert boundary for both 1972 and 1979 was in roughly the same position as Lamprey's 1975 line.

ii) *Problems of monitoring vegetation shift*

The above example encapsulates many of the controversies of the entire desertification debate. First, the simple fact that a single study has been used to quantify the global problem of desertification and secondly, that even this study may not have been accurate.

The Swedish team's findings are perhaps best taken as an illustration of how difficult it is to assess and monitor the degradation of vegetation, rather than as absolute proof that there has been no shift in vegetation belts. Similarly, the interpretation of causes behind observed changes in vegetation belts are complicated by such factors as interannual variations in land use, rainfall, disease and insect infestations, the effectiveness of burning practices and availability of water supplies.

Figure 2.10 illustrates the very great changes in biomass production that can occur in dry

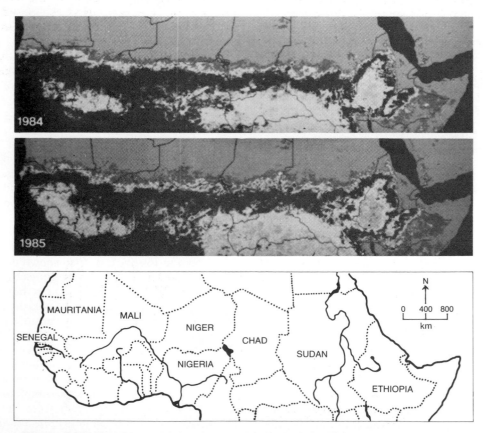

Figure 2.10 *Vegetation fluctuations across the Sahel, 1984 and 1985 wet seasons*

regions. The images are based on satellite information for the wet seasons of 1984 and 1985 across the south side of the Sahara. The arid/semi-arid boundary is the first change of shade from grey to the north (Sahara Desert) to a darker grey. In 1984 this limit is almost coincident with Mauritania's southern border, while in 1985 the limit has progressed into Mauritania's southernmost zone. Along the Sudan-Chad border the arid/semi-arid boundary has shifted about 200 km and about 150 km along the Sudan-Ethiopia border between 1984 and 1985.

1984 was the driest year in at least thirty years, whereas 1985 was the wettest year since 1981. 1984 was a year of widespread crop failures, whereas 1985 saw nearer average rain-fed grain production in Senegal and Niger and the best yields for several years in Sudan, Mauritania, Mali, and Chad.

Regional differences are also marked. In the El Obeid area, south of Lamprey's study zone, the 1984–1985 boundary shift was about 130 km; 200 km west of El Obeid, the northward shift was just 50 km.

These satellite images show that interannual shifts of vegetation belts of 50 to 250 km are normal in desert marginal areas. Since this is the case, a permanent shift in the vegetation belts of between five and six kilometres, as suggested by Lamprey, should need between 30 and 40 years of observation from meteorological satellites and ground studies to decide whether such a shift was permanent or not. Since desertification is patchy in its development, a still longer time would be needed to determine whether the changes were temporary or permanent.

To date, such extended studies are still in their infancy. If satellite observations are to be used for monitoring desertification, and they appear to be a very useful tool for monitoring on a broad scale, then many more years of data are needed before any satisfactory results can be achieved.

iii) Dust storms

Just as meteorological data on the frequency of dust storms has been used as an indicator of desertification in Mauritania, so these events can be monitored in Sudan.

Figure 2.11 shows the variation in annual dust storm frequency and annual rainfall totals for El

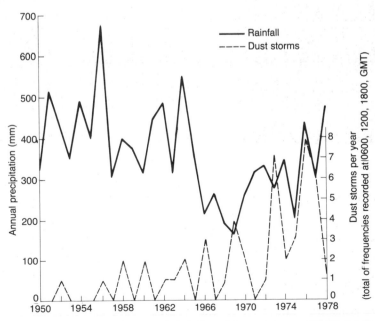

Figure 2.11 *Annual variation in rainfall and dust storms at El Obeid, Sudan*

Obeid in central Sudan. A marked rise in dust storm activity can be seen clearly dating from the late 1960s/early 1970s. Similar patterns can be seen in the data from other Sudanese stations across the Sahel such as El Fasher, Khartoum and Tokar.

iv) Has desertification occurred?

In the case of El Obeid, the team of Swedish geographers who monitored central Sudan from satellite and field surveys found no evidence of vegetation loss or desertification in the El Obeid region during the period 1964 to 1974. Their findings are at odds with the meteorological evidence on dust storms.

Another study in the area just west of El Obeid looked at some further aspects of desertification. A Sudanese team studied the plants of the savannah grazing lands of the Kordofan Region over the period 1963 to 1977. They looked at the percentages of plant cover, vegetation litter and bare soil over the period and also the composition of the vegetation itself, measuring the diversity of plant species and those parts of the total vegetation cover made up by particular species.

The results from their three test sites showed a substantial and steady decrease in the number of plant species. At one site, for example, a total of 17 different plant species found on the open range in 1963–65 had fallen to just four species by 1971–77. The composition of species types had also changed, so that in 1971–77 there were no *perennial* (i.e. plants that live for several years) grass species left. Such a loss may have been a result of either overgrazing or drought, but the increase in unpalatable perennial herbs at the site was considered to be a result of selective grazing.

All sites also recorded a decline in plant cover and an increase of bare soil. The two sites close to centres of population showed the greatest changes, suggesting the effects of higher rates of fuelwood collection and grazing.

vi) Summary of the evidence

The situation in Sudan is clearly complex. It is perhaps one of the best-studied parts of the world for desertification and yet the evidence is conflicting. Lamprey's study has been criticised by the Swedish team. Yet evidence from dust storm observations and monitoring of vegetation indicates that some form of desertification has occurred. The situation here serves to illustrate the problems of monitoring desertification.

2   DESERTIFICATION OUTSIDE THE SAHEL

2.1 Kuwait

Kuwait, one of the richest countries in the world in terms of per capita income, starkly contrasts with the Sahelian countries which are among the poorest (Figure 2.12). Yet Kuwait also suffers from the problem of desertification.

In this case there is little doubt that desertification problems have been caused by human action as opposed to climatic factors. The rapid rate of development and population growth (Figure 2.13) which followed the country's first oil exports has increased the human impact on Kuwait's environment and caused the problems of desertification. Kuwait also provides an example of desertification happening not at the desert margin but within the desert itself.

i) Physical geography of Kuwait

Kuwait's climate is either arid or extremely arid. Annual average rainfall is about 100 mm a year and the mean daily evaporation rate is 16.6 mm per day. Temperatures are high: the mean July temperature is 37.4°C and the maximum reaches 45°C.

Summer is also the time of highest wind speeds, which blow predominantly from the north-west. These winds cause two other typical features of the climate of Kuwait: sand storms and dust storms. Sand storms are blown from the sand deposits mainly in the south of the country, while dust storm sediments are predominantly blown from the dry muddy floodplains of the Rivers Tigris and Euphrates in Iraq. A dust-laden

Before the first exports of oil in 1946, Kuwait's economy depended mainly on pearl diving, sea trading and fishing. The population in the 1940s was small, less than 150 000, and the people were largely found in old Kuwait City and small agricultural settlements, while some herders were able to make a living in the harsh desert environment.

Since oil has taken over as the main driving force behind Kuwait's economy the country has experienced dramatic changes in its society. Population has grown rapidly to 2.1 million in 1989. Much of this population lives in urban areas. Today the whole coastal area from Kuwait Bay to the country's southern border is an urban strip about 10 km wide.

Such rapid development has increased the human impact on Kuwait's environment and caused the problems of desertification.

Figure 2.12 Kuwait in context

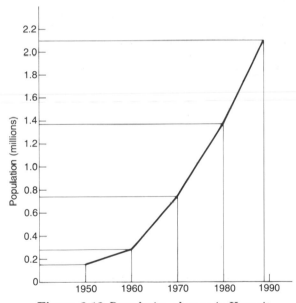

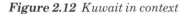

Figure 2.13 Population change in Kuwait

atmosphere is such a common feature of Kuwait's climate that it occurs on almost 80 per cent of days each year. Sometimes the dust is so thick that the visibility is reduced to just a few metres.

Kuwait's land surface is flat and covered with deposits of sands and gravels with numerous dry salt lakes or *playas* (Figure 2.14). The country's vegetation is sparse and is made up largely of small shrubs.

ii) Evidence of desertification

A desertification problem in Kuwait has been recognised in the last 30 years. Before the 1960s the country's terrain had, for the most part, wind-stable surfaces. In some areas these surfaces were covered in sparse vegetation which protected the finer material around it from wind action. In other, non-vegetated, areas the wind had removed all the fine-grained material to leave a gravelly surface too heavy to be moved by all but the very strongest gusts.

In recent years, however, Kuwait has experienced increasing problems from sand, with a considerable increase in sand movement and accumulation, and a noticeable reduction in the density of vegetation cover. Observations of these phenomena in the field can be detailed under a number of headings.

a) Loss of vegetation from rugged sand sheets

Sand sheets, protected from wind action by a rugged vegetation of desert shrubs, were formerly common features of the wide shallow depressions surrounding playas in the western part of the country, and infilling wadi systems. These sand sheets were immobile, forming domed accumulations around individual shrubs.

Today, however, these features are hardly visible, apart from at a few locations in the northwest. In their place are smooth sand sheets virtually bare of vegetation. These features have been transformed due to a loss of vegetation.

When the shrubs are removed the soil surface is no longer protected from the wind. The organic-rich topsoil is also deflated, leaving a surface of coarse sand and gravel. The landscape now consists of rippled sand sheets and gravel ripples.

b) Deflation of playa mud

Playas are small, often salty, dry depressions, which range in area from 1 to 4 km². Surveys in 1980 showed that playa surfaces were covered in mud and sandy mud, with very flat well-structured soil supporting abundant perennial vegetation. The areas around playas were usually covered by large sand sheets.

During the 1980s these desert depressions have changed. In the southern desert, where the playas are mostly small, they have been engulfed by moving sand, and the surrounding rugged sand sheets have lost their plant cover and their sands.

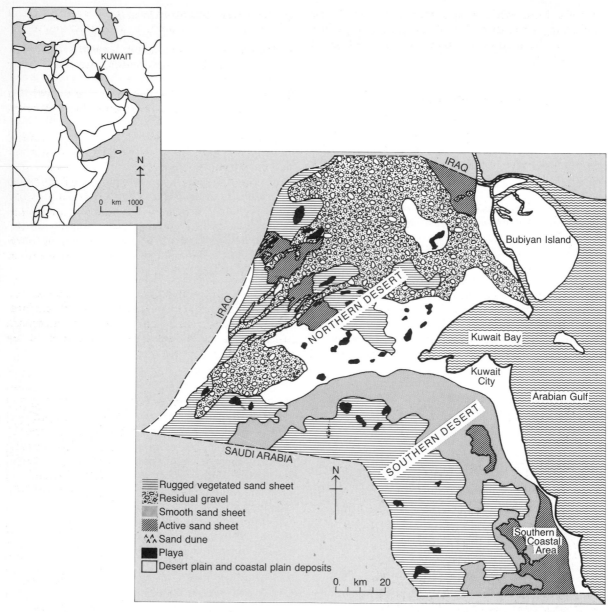

Figure 2.14 *Surface sediments in Kuwait affected by desertification*

The largest playas in the northern regions have been severely eroded, losing up to 0.7 m of mud from their surfaces. Where mud remains on the surface, these more compact areas have been carved by the action of the sand-laden wind into *striations*, aligned with the northwesterly winds.

c) New barchan dunes

During the second half of the 1980s new dunes have formed in an area of northern Kuwait that in 1980 was a zone of gravelly deposits. In a period of five years small *barchan* dunes have been built up. A barchan is a crescent-shaped mobile form with a slip face and two horns pointing in the direction of the prevailing wind. Some of these dunes are now four metres high at their crests.

iii) Causes of desertification

Kuwait's rapid development and growth, financed by its oil wealth, has resulted in a number of increased impacts on the desert ecosystem. These causes of desertification can be outlined under a number of categories.

a) Sand and gravel quarries

The rapid growth of Kuwait's population, most of which live in cities, has meant a boom in the country's construction industry. The demand for aggregate materials such as sand and gravel, for the production of concrete, road pavement and coastal area reclamation, has exploded, while limestone deposits and calcretic sand have been used for fill material. All these materials are found in shallow deposits and are quarried using open-cast techniques (Figure 2.15). Gravel

quarries have produced the worst environmental effects. The main problems are shown below.

1. Destruction of vegetation and loss of associated wildlife on production sites and their surrounds.
2. Removal of gravel which protected underlying finer deposits, thus exposing them to wind action.
3. Accumulation of huge spoil heaps around quarries, usually made of finer sand and silt dug up with the gravel. These loose, fine grained tailings have become significant new sources of wind-blown dust and mobile sand.
4. Release of large amounts of dust particles into the air during sieving and crushing of the quarried gravel. These clouds, which often reduce visibility to a few metres, are common sights around quarries.

The accumulated effects of these quarries has been to destabilise the natural desert surface. This has destroyed vegetation and increased erosion of sand and dust, classic symptoms of the desertification process.

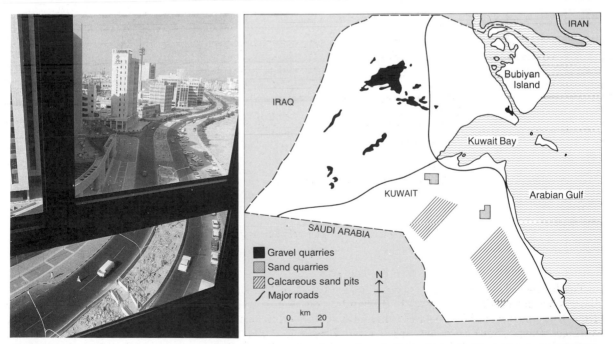

Figure 2.15 *The map shows the distribution of sand and gravel quarries in Kuwait. These have supplied the country's booming construction industry*

b) Off-road traffic

The movement of vehicles off paved roads has also increased in the last 20 years or so. Heavy construction machinery, and lorries transporting building materials, are perhaps the worst offenders, but personal cars and motorcycles add to the problem around the cities.

This uncontrolled traffic has destroyed plants and compacted ground surfaces, so making them less suitable for plant regrowth. Vehicles also generate a good deal of atmospheric dust.

c) Overgrazing

Kuwait's vegetation cover has also suffered from the grazing of too many livestock, often beyond the land's carrying capacity. When plants are lost the soil is then susceptible to deflation and in many areas the result is mobile sands. Woody desert shrubs have also been cut down and uprooted to be used as fuel.

iv) Desertification hazards

Although wind erosion has long been a typical and common feature of Kuwait's environment, the recent worsening of conditions by human factors has increased the occurrence of hazards associated with *aeolian* or wind-generated processes.

Reduced visibility during dust storms is a hazard to transport, affecting the frequency of take-offs from airports, the movement of shipping and motor transport. Breathing problems, eye infections and allergies are increased when dust is airborne.

Sand movement and build up causes problems by blasting buildings, oil installations and pipelines and accumulating around obstacles. Pipelines engulfed by sand tend to corrode more quickly, oil drilling and pumping operations are often suspended during sand and dust storms, and accumulations are expensive and time-consuming to remove.

New problems have been caused by applying inappropriate measures to control the movement of sand. In the case of roads, for example, the solution was to construct parallel earth walls to give protection. Unfortunately the earth barricades only made the problem worse since sand accumulated on the slopes and formed an elongated tongue which stretched across roads, eventually blocking them. Subsequently, the earth walls have been removed and the sand allowed to blow freely across the highways.

v) Combating desertification

Unlike many other world areas suffering the effects of desertification, the problems in Kuwait are undoubtedly caused by human activities. Solutions to the desertification problem should therefore be relatively straightforward, since, as explained in Part I, suitable control measures depend upon accurate identification of causes.

Kuwait is fortunate in having adequate financial resources to put plans for combating desertification into action. Suggestions include: the conservation of areas not yet desertified through land-use designation; zones could be delimited for particular usage such as campsites, for vehicle use only or as national parks; rehabilitation of desertified areas, to reduce the hazards of mobile sand; stabilization of sand sheets, by reseeding and replanting.

1 BACKGROUND TO THE DESERTIFICATION PROBLEM IN THE SAHEL

1.1 The development context

The proportion of a country's population who are directly affected by desertification will depend upon how many people are employed in the agricultural sector. Table 3.1 shows that the Sahelian countries listed have very large proportions of agriculturalists in their populations. Column B gives an indication of a country's wealth, which gives some idea of how able a country is to pay for desertification control.

i) These countries are in no particular order. In order to decide whether there is any relationship between a country's wealth (Column B) and the proportion of its labour force in agriculture (Column C) perform a Spearman's Rank Correlation calculation on the data.

Spearman's Rank test gives a value which shows the degree of association between two variables. The value ranges from +1.0, through 0.0, to −1.0. Zero indicates that there is no association between the variables, while 1.0 indicates a perfect correlation. The + or − indicates the direction of the association. The following steps should be taken.

a) State the null and alternative hypotheses.

Null hypothesis: There is no significant association between GNP per capita and percentage of workers in agriculture.

Alternative hypothesis: There is a significant association between GNP per capita and percentage of workers in agriculture.

b) Convert the data from the original ratio scale to the ordinal scale by ranking the countries according to their GNP per capita from highest (rank 1) to lowest (rank 10). Enter the rank in the table at Column D.

Table 3.1 GNP per capita and percentage of labour force in agriculture

A Country	B GNP per capita (US$)	C % labour force in agriculture	D rank B	E rank C	D−E	(D−E)²
Niger*	260	91				
USA	17 480	4				
Morocco	590	46				
Colombia	1230	34				
Sudan*	320	71				
Greece	3680	31				
Poland	2070	29				
Mauritania*	420	69				
Spain	4860	17				
New Zealand	7460	11			Total (D−E)²:	

* Sahel

c) Convert the data to rank values in the same way for percentage labour force in agriculture and enter under Column E.

d) Calculate the difference (D−E) for each country and then square the result.

e) Sum the values for (D−E)^2 and enter into the equation given below.

Spearman's Rank Correlation Coefficient, r_s

$$r_s = 1 - \left(\frac{6 \times \sum\limits_{i=1}^{n} d^2}{(n^3 - n)} \right)$$

where d = the difference in rank between two data sets (i.e. D−E)

n = number of paired observations

ii) The resulting value of r_s needs to be tested for significance. The Student's t test should be used for this. The formula is as follows

$$t = \sqrt{\frac{n-2}{1 - r_s^2}}$$

The tabulated value of t at the 95% significance level, with 9 degrees of freedom (i.e. number of pairs − 1) is 2.26. If the calculated value of t exceeds the tabulated value the results are significant at the given level.

iii) What does the correlation value tell us about the relationship between a country's GNP per capita and its agricultural labour force?

iv) From the tabulated data suggest *three* problems facing Sahelian countries in their attempts to overcome desertification.

2   MAURITANIA

2.1 Mauritania – priorities for development

The following is a role playing exercise.

The scene is a consultative meeting in Nouakchott to discuss the desertification problem, how it can be tackled, and where it should come in a list of priorities for Mauritania's development to be adopted in the next five year development plan for the country. The characters in the meeting are outlined below.

The characters

Abdullah Mahmoud is Industry Minister and chairs the meeting. He has been the Minister since 1982 and has overseen a decline in Mauritania's mining exports (mostly of iron ore) since the 1970s as iron ore prices have declined. He firmly believes that industrial development is the way forward for Mauritania and that further development of ore processing industries is necessary. Mauritania has iron ore reserves of over 3 billion tonnes, the fourth largest in Africa. He believes that problems of desertification are primarily caused by the drought and that this problem will not go away. Hence his commitment to industrial development is strengthened.

Abdullah is 42 and is an ambitious politician. He has a military background and he has little patience with undisciplined thought. He is used to giving orders, and believes that women should not be involved in high-level decision making. He has little formal education and is prone to dismiss it as unnecessary, however, he does have a great deal of respect for Abdi Riyadh's ideas and advice. Although his grandparents were farmers in the Senegal River Valley, he has worked to distance himself from what he sees as a poor background. He firmly believes that the way to improve the poor farmers' lot is to industrialise.

Omar Durand is Minister for Fisheries and Agriculture. He thinks that the fisheries industry, which has rapidly climbed to become Mauritania's biggest export-earner during the 1980s, is the country's key resource that should be further developed. He wants to expand the fishing fleets and introduce more mechanised ships. His interest in agriculture is confined to the fertile Senegal River alluvium. In his opinion the nomadic pastoralists represent an outmoded method of land use.

Omar is 54. He is of Mauritanian father and French mother. He was educated in France and enjoys a Europeanised lifestyle.

Michelle Heeb is an International Red Cross worker. She spends most of her time administering the refugee camps around Nouakchott. She takes a strong *hands on* approach, involving herself with the day-to-day running of the camps and dealing with the problems of malnourishment and displacement. She is an experienced Red Cross worker, having worked in other parts of the Sahel and in the French-speaking Far East. She is constantly badgering the national government to put more resources into the refugee camps and believes that the country can only get back onto its feet when the plight of the refugees is sorted out.

Michelle, 45, has long experience of dealing with developing country diplomats and civil servants. She stands her ground well and never loses her cool.

Brenda Morley is the Mauritanian representative for the US Agency for International Development (USAID). She believes in helping the drought victims as much as possible and she is aware that there is surplus US grain that can be offered in food-for-work programmes among the refugees. She also wants to encourage development of the fishing industry, but at the scale of the individual fisherman.

Brenda is on her first assignment for USAID. She is 29, has never been to Africa before, but has spent a short time working with a development project in Thailand. In her own mind she is not totally convinced about food-for-work programmes since she has read that they tend to undermine local food prices and thus may discourage people from growing their own food crops. She cannot speak French and thus needs to work through an interpreter.

Mitsui Ishikawa is a representative from the Japanese government's overseas aid programme. He has a big budget to play with and his brief is to secure rights to Mauritania's raw materials: fish and iron ore. He is not interested in environmental matters, but is aware that some money should go towards technical aid for refugee camps.

Mitsui is 42. This is his first visit to Africa and he does not like it. He also needs an interpreter, but the others around the table are more willing to listen to him than to Brenda Morley because they know that he probably has a large budget. Privately, he has spoken to Abdullah Mahmoud and they have agreed that Japanese iron and steel experts will visit the country to advise on which direction the country's industry should progress.

Abdi Riyadh is a Professor in Development Studies from the National University. He has just begun to advise the Mauritanian government on appropriate policy for economic development. Having seen how dependent Mauritania has become upon iron ore as a source of foreign income he is a strong believer in as broad a base as possible for his country's economic development. He is particularly supportive of policies for sustainable development (the idea that countries should only develop using their resources in a sustainable way) and likes to look at things in the long term. He thus supports plans for expansion of the fishing industry and realises that iron ore will not last for ever. His view of the Sahel problem is that desertification is essentially a problem brought about by people. Thus, he firmly believes that the problem can be overcome with the appropriate policies.

Abdi is 45. He studied in France and the USA before returning home to Mauritania. He is familiar with politics and high-level negotiations since he has worked on broad development issues for the United Nations Environment Programme in Nairobi. His true feelings about his country's politics are that the military have been in charge much too long. He tends to oppose anything they suggest, but with very well-argued objections.

Notes: Since Brenda Morley and Mitsui Ishikawa can only speak through interpreters, two more students should act as interpreters, repeating anything they are told by these two characters.

i) Each student should pick a character and prepare a five minute talk/presentation, giving a) their plan to tackle the desertification problem, and b) their priorities for development over the next five years.

ii) Having heard the presentations, Mr Mahmoud will then open the floor for discussion.

iii) After the meeting each character should do the following:

a) Write an official report on the meeting to your superior.

b) Write a covering letter to go with the report which expresses your real feelings about the goings on at the meeting.

iv) Now change role. Assume you are a journalist covering the meeting and that your brief is to file a story on Mauritania's plans for dealing with the desertification problem. Write your story in no more than 500 words. Once you have decided what your rough storyline is, you may like to arrange a private interview with one or more of the participants. Remember that you can quote individuals but you should not libel anyone.

3 SUDAN

Country blowing away across the Atlantic

By Nick Middleton

The West African state of Mauritania is blowing away. Every year 100 million tonnes of topsoil, much of it from the fragile Sahel zone, is sucked up by the burning desert winds and carried out over the North Atlantic.

The huge quantities of dust have been seen clearly on satellite images, billowing over the sea; a great tongue of soil that stretches as far as the Caribbean and South America.

The latest data shows conditions have worsened in the two years since a report in *Nature* described the growing storms.

The reasons for this ecological disaster are well-known: drought and desertification, twin horsemen of the apocalypse familiar to the inhabitants of the Sahel region of North Africa that extends from Mauritania in the west to Sudan and Ethiopia in the east.

Mauritania, a country roughly the size of France and Spain combined, is largely Sahara Desert, with a more hospitable zone in the south, the Sahel. The desert regions are strewn with moving dunes: sand seas or *ergs* in the Arabic. Vegetation is sparse and wind is the dominant agent that sculptures the landscape. But this wind eroded region has expanded southwards into the desert-marginal Sahel zone since the late 1960s.

Overgrazing, overstocking and cutting woody shrubs for fuel as world oil prices rose, coupled with a population explosion, all spelt disaster when drought came.

The resulting desertification, like an advancing tide of sand, is being monitored by scientists at Nasa headquarters in the United States. They have watched the gradual development of a strip of new desert across the Sahel.

Since the 1960s the dust storms in the Sahel of Mauritania have been increasing. Nouakchott, the capital, usually had five severe storms a year. Since 1983 dust storms have blown on more than 80 days each year.

The storms often advance as a wall of dust, hundreds of metres high, engulfing all in their path, reducing visibility to near zero. The dust gets in your eyes, up your nose, through doors and windows, coating everything in a thick layer of red dust.

Cars and airports are brought to a standstill, radio communications are affected. But worst of all, the soil is lost, piling up on the bed of the Atlantic Ocean.

With the Sahel's vegetation destroyed and soils disappearing the pastoralists have flocked to the cities. In 1960 Nouakchott had a population of less than 20,000, today it has 350,000.

Malnutrition affects one child in three. The average life expectancy is 46.

But the possible consequences of Mauritania's ecological disaster do not stop there. Dust remaining in the atmosphere over the country could be helping to prolong the drought.

There is discussion among climatologists as to whether the current Sahel drought reflects a change in climate, or is a temporary fluctuation. Certainly the Sahel has experienced changes in its climate.

It was much drier than it is today. During this dry period the present Sahelian area of Mauritania was a mass of moving dunes. Since then the climate has become milder and these dunes have become stabilized as grasses and scrub grew on their slopes.

With dust storms blowing away the thin soil, the dunes are beginning to move once more.

Once mobile, it is possible to stop sand dunes, but it is not easy. Even in oil-rich Saudi Arabia the techniques used are not always successful.

Figure 3.1 *Extract from The Times, September 1987*

3.1 Sudan's desert crisis

Figure 3.1 shows a copy of an article that appeared in The Times in 1987. You will see that it uses much of the information on Mauritania that has been discussed in Part II.

i) Assume you are a journalist covering the Sahel drought and its consequences. Write a similar piece not more than 600 words long, the next in a series, on the situation in Sudan using the information in Part II. You may be allowed an illustration by your editor. Suggest the sort of subject matter you think appropriate photographs should show to support your article.

3.2 Drought and the farmers' response

Table 3.2 shows some data for an area of intensive cultivation near El Obeid in Kordofan. Millet yields are given in feddans, the unit of field measurement used in this part of Sudan. One feddan is about 0.42 hectares.

i) Plot the data for millet yields against time, with the year on the x axis. Describe your curve.

ii) Compare your curve with the curve for rainfall at El Obeid shown in Figure 2.11.

Comment on the relationship between rainfall and millet yields.

iii) Now plot the data for annual millet production against time on a separate graph, with the year on the x axis. Describe your curve. How do you think this increased production can have occurred?

iv) Calculate the area cultivated each year in thousands of feddans. Note that whereas yield is given in kilograms the annual production is in tonnes, so you may have to convert one of the units.

v) Plot the data for area cultivated against time on another graph, with the year on the x axis.

vi) Referring to your graphs, suggest how farmers near El Obeid may have responded to declining millet yields during a drought period.

vii) What other factor, besides a lack of annual rainfall, could have contributed to declining millet yields during the late 1960s? How will the farmers' response to drought affect this other factor?

viii) In order to prove that desertification is or is not occurring in this area, what other data sets would you need? How many years of data do you think would be necessary?

Table 3.2 Millet yields and production in an area near El Obeid, Sudan

Year	Millet yield (kg/feddan)	Millet production (thousand tonnes)
1961	240	80
1962	240	80
1963	220	80
1964	205	90
1965	205	100
1966	190	100
1967	175	100
1968	190	110
1969	190	125
1970	190	145
1971	145	130
1972	80	120
1973	65	115
1974	80	115
1975	95	125
1976	120	150
1977	130	170
1978	175	210

3.3 Vegetation changes in Sudan

Tables 3.3, 3.4 and 3.5 show the data taken from a study of vegetation change in the grazing lands of Kordofan Region in Sudan (Figure 3.2).

i) Assume that you are employed by the United Nations Environment Programme working as part of a new project they have set up to educate and increase awareness of the desertification problem among desert nomads in Sudan.

Using the results of this study, devise a scheme to get the information across to groups of desert dwellers. For the sake of this exercise, you can assume that your audience can read, but not very well. Hence you should rely on pictures and simple diagrams as much as possible and keep text to a minimum. Your budget is limited, so your scheme can consist only of a short slide show and a permanent display which you can transport around with you. The display consists of three poster boards, each two metres by two metres.

a) What would you show on the boards and what would your slide show consist of?
b) Prepare a mock-up of one of your display boards.

The following account of an aid worker in central Africa will help to illustrate the sorts of conceptual problems that your audience might experience and that you should be aware of.

The aid worker was charged with touring a part of central Africa with a short film about the dangers of malaria and appropriate ways in which it can be avoided. The makers of the film, wanting to bring home the real dangers of the apparently harmless little mosquito, emphasized the point by including some close-up shots of the nasty-looking insect so that it filled the screen. This shot caused widespread unrest among the audience, many of whom had never seen a film before. But during the discussion afterwards the aid worker was disappointed by the lack of alarm and interest in his audience at the dangers of mosquitoes and malaria. When asked why not, members of the village replied that there was no problem in their region because their mosquitoes were tiny, not the size of elephants as in the unfortunate region portrayed in the film.

Table 3.3 Species diversity

	1963–1965		1971–1977	
	OR	EX	OR	EX
En Nahud				
Perennial grasses	3	2	–	–
Annual grasses	3	3	1	3
Perennial forbs	1	1	2	–
Annual forbs	10	10	1	1
Total	17	16	4	4
El Khwuie				
Perennial grasses	2	2	–	–
Annual grasses	4	8	2	3
Perennial forbs	–	3	1	2
Annual forbs	11	12	2	1
Total	17	24	5	6

Note: forbs are grazing herbs other than grass
OR: Open Range   EX: Exclosure

Table 3.4 Percentage botanical composition

	1963–1965		1971–1977	
	OR	EX	OR	EX
En Nahud				
Perennial grasses	19.3	11.3	0.0	0.0
Annual grasses	18.4	27.1	46.8	83.0
Perennial forbs	0.1	2.0	34.8	–
Annual forbs	62.2	39.9	18.4	17.0
El Khuwie				
Perennial grasses	3.6	4.4	0.0	0.0
Annual grasses	11.7	24.0	16.9	26.4
Perennial forbs	0.0	1.6	20.0	12.0
Annual forbs	85.1	72.2	63.1	61.6

OR: Open Range   EX: Exclosure

Table 3.5 Percentages of plant cover, litter and bare soil

	1963–1965		1971–1977	
	OR	EX	OR	EX
En Nahud				
Average plant cover	20	14	14	38
Average litter	11	19	1	18
Average bare soil	69	57	85	44
El Khuwie				
Average plant cover	25	31	15	30
Average litter	21	22	1	12
Average bare soil	54	47	84	58

OR: Open Range   EX: Exclosure

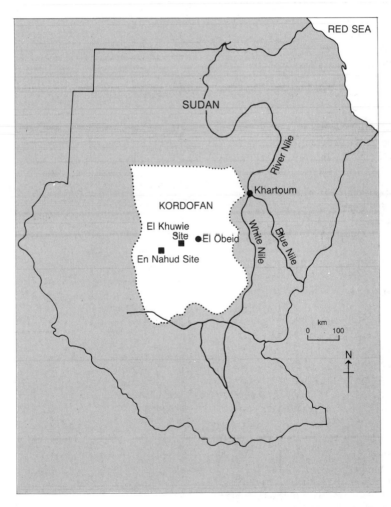

Figure 3.2 *Location of vegetation monitoring sites, Kordofan, Sudan*

SUGGESTED READING

Heathcote, R.L. (1983) *The arid lands: their use and abuse.* Longman.

United Nations (1977) *Desertification: its causes and consequences.* Pergamon.

Warren, A. (1984) The problems of desertification. In Cloudsley-Thompson, J.L. (ed) *Sahara desert.* Pergamon.

Worster, D. (1979) *Dust Dowl.* OUP.

INDEX